C000126821

ISSUE 36:
AUTUMN 2023

ISSN: 2059-2590
ISBN: 978-1-7395359-0-2

Submissions of fiction, art, reviews, poetry, non-fiction are welcomed: visit the website to find out how to submit

www.shorelineofinfinity.com

Publisher
Shoreline of Infinity Publications /
The New Curiosity Shop
Edinburgh
Scotland

300923

Guest Editor:
Lyndsey Croal

EDITORIAL TEAM

Co-founders:
Noel Chidwick,
Mark Toner

Fiction Editor:
Eris Young

Reviews Editor:
Ann Landmann

Non-fiction Editor:
Pippa Goldschmidt

**Marketing &
Publicity Editor:**
Yasmin Kanaan

Production Editors:
Noel Chidwick,
Andrew Lyndsay

Copy-editors:
Pippa Goldschmidt
Russell Jones
Iain Maloney
Eris Young
Cat Hellisen

CONTENTS

- - - - - - - - - - -

COVER ART
Vincent Kings

FIRST CONTACT
www.shorelineofinfinity.com
contact@shorelineofinfinity.com
Twitter: @shoreinf
Also on Instagram & Bluesky

www.shorelineofinfinity.com/36-supplemental/

PULL UP A LOG

Guest Editor Lyndsey Croal

I'm delighted to introduce Shoreline of Infinity's Special Climate Change Issue. As an author, science fiction fan, and climate change policy professional, this has been a dream project and a huge privilege to work on. I've been blown away by the quality of work we received and are publishing in this issue, as well as by the support during our Kickstarter – thank you to all of you who helped make the issue happen.

Climate change is likely to be the biggest challenge we face in the coming years. Exploring climate and eco themes in science fiction gives us an opportunity to explore not just the consequences of climate change, but also allows us to imagine the various alternative futures that could open up if we take real concerted action. Fiction is a powerful tool for connecting us with these potential visions for change, and as Kim Stanley Robinson says in our interview with him in this issue, "fiction leaves traces in our mind" – a powerful notion that what we read stays with us and can inspire action.

Contained in this special issue you'll find diverse visions and reflections on climate change from around the world, exploring its impacts, and how we might address future and current challenges. Some of the work envisions action on a grand scale while others propose more community-centred approaches. While a few are set in further-off, dystopian futures, they still offer an insight into how the world might rebuild and learn lessons from the past.

Throughout the issue, there will be something for every reader – thought-provoking essays and reviews, stunning artwork, and poetry and stories full of action and adventure, fable and folklore, Solarpunk visions, far-flung futures, and cosy settings. There are themes throughout of family, friendship, and love, of cherishing our natural world, with many stories exploring our relationship with flora, fauna, fungus, and even spiders, that might make you think twice about our connections to nature. There's also exploration of new tech and artificial intelligence, with robotics that surprise, and ethical questions to consider and explore.

All of the pieces in this issue are special in different ways – they are important and powerful, and allow us to imagine solutions, consequences, and the sort of communities we could be living in if we truly faced up to the climate crisis. I hope that this issue will entertain and inspire, as well as showcase the variety of narratives science fiction can offer us in thinking about some of the most important issues of the day.

I'll not keep you any longer – happy reading, and thank you again for supporting Shoreline of Infinity.

red and green, orange and blue, me and you

Kelley Tai

A note from the author:

It's monsoon season in Taiwan, and monsoon season means more rain, more humidity — and more typhoons.

Taiwan is a small island off China's east coast, west of Japan, and north of the Philippines, about half Scotland's size. And every year, three to five typhoons hit Taiwan.

Typhoons are tropical hurricanes powered by hot air rising from the ocean's surface and cool air falling back down. However, 2023 is an El Niño year. This means typhoons are formed earlier and further away from Taiwan. As these typhoons travel longer distances, they're charging toward the island with more energy. Last month, Typhoon Khanun hit Taiwan. In Taipei, the capital of Taiwan, businesses shut, schools closed, and soldiers were on standby. Massive landslides destroyed roads in Nantou County. The indigenous communities living in the mountains had to be airlifted to safety. What is interesting about typhoons is that a decrease in them is equally as horrific. If too few typhoons hit Taiwan, the country could be caught up in a drought because of a decline in rainfall.

Climate is delicate — one small change will tip the balance, causing irrevocable damage. Typhoons need to hit Taiwan, but it needs to remain the same amount, with the same level of intensity. With collective action and technology, we can keep our typhoons at bay, and also create a more sustainable future — full of empathy and reciprocal love for the planet.

Today is the Country's 100th anniversary, and instead of getting drunk off gin figg and toasting to the giant balloon of our national plant (a very lovely Hedgehog Cactus), I'm stripping off my clothes in an oasis in the middle of the desert.

I feel like I'm inside one of Monet's impressions. Girl, naked; world, a play on light.

Our desert is a painter's dream, beautiful in pairs of color theory: red of the charred canyons, steeping the hydroponic farms evergreen; blue of Lake Qing, saturating the sun when it sets low in orange, not as a shade, but a celebration of another day.

Hear, hear!

I fling my undergarments onto the pebbled shore and watch them fly.

Over the lake, a red-tail hawk *keeeeeearrs*. Yéye used to say a red-tail's song brings power to the desert, so I'm taking her cry as my personal blessing today.

Another red-tail glides by. But this time, the CondensAir™ drone's low hum gives itself away.

The bird gives the imposter a curious look.

Grandpa taught me the names of all volant animals using these machines that turn air into water. The drones are designed to look like falcons, ospreys, even dragonflies and hummingbirds, to be harmonious with nature.

The mist from the CondensAir™ drips down from my face to my collarbones, down the valley of my chest, leaving me with chills in the afternoon heat.

I bet the cattails like it, too as they sway on the lake's edge, dancing like no one's watching—

Boom.

Boom.

Boom.

In the distance, the heavy cadence of the percussion line crescendos with the procession in a chorus. My bones pulse to the rhythm of our anthem. *'We are one. Love for the Desert! Love for Each Other!'*

I look to the Capitol and immediately regret it. 'Ahh, shit.'

When our world leaders told us our future would be bright, they meant it literally, with solar panels plastered on every roof, every wall, every sky train. A mirage happens when light bends through layers of air at different temperatures. Sometimes, I can't tell if the Capitol is the thing bending those lights or if it is a mirage itself.

I blink in and out, out and in, and when the white spots finally fade, I see the red-tail's folded wings, her eyes locked onto a leopard frog.

The Capitol can go on without me. Grandpa and I never minded missing out on the parade. We celebrated more festively anyway *and* made it back in time to watch the fireworks. A win and an extra win. Out here there are no parties, after-parties, or loud noises; just me, the desert and—

'CALL FROM OPAL.'

I jump back at the sound, a cry buried in my chest.

The Computer Representative Operational Wizard, or CROW, my AI assistant, says, 'CALL FROM OPAL. CALL FROM OPAL.'

Ring.

Ring.

Ring.

The notifications come from my in-ear headphones. The familiar bells and alerts of the call's ringtone feel out of place here, too shrill for the tranquility of the Capitol's outskirts, near our once-sacred oasis. But mostly... because the call is from Opal?

I don't know any other Opals besides the one from uni. I mean, it *is* the Opal from uni.

Her caller ID hovers in front of my eyes, blurring out the

desert behind, a name in bright neon blue: OPAL VAN DER LOPEZ.

I fumble in disbelief as I read her name over and over.

Should I pick up? What would I say? I already told her I was busy when she asked me about my plans for Country Day.

I let it ring all the way.

Once it's quiet, I breathe out a sigh, the lake returning into view.

'New video message from Opal,' CROW says. 'Playing new video message from Opal.'

'Wait—'

Too late.

'Hey, Tian.'

She looks so much like she's here with me. Or, I look so much like I'm at the school with her. Opal wears a flower crown on her head, as most citizens do on Country Day. She's on the garden rooftop, where we've been experimenting with different types of desert soil best for farming. I can see our handwritten labels taped on big pots of loam. My words look like chicken scratch compared to hers. I don't know why she never believes me when I tell her.

Opal clears her throat. 'It's Opal...your partner from the pedology course. Obviously.' Her monarch butterfly earrings flutter as she shifts her weight from one foot to another. 'Umm.....I know you said you've helped out the arfajips with your Grandpa before, but I wanted to just call and say... stay safe. I know people say it's archaic, but I think it's cool, what you're doing. I mean I could never. Anyway, I know you already said no, but feel free to come join us after—'

I hurl both hands to the right, and Opal's face disappears.

The red-tail is gone, and so is the frog on the lilypad.

'CROW,' I growl, 'Turn all notifications off.'

'All notifications off,' CROW replies.

I strap the oxygen tank on my chest into place. Maybe I should disconnect from CROW completely. In every other circumstance, I would, but if things get bad under the lake, that 1% chance of something going wrong, especially with Yéye no longer watching over me…

The last thing I need is still in my knapsack.

I crouch down on my knees and take out Grandpa's weighted belt.

Yéye gave me his belt the year he stopped diving with me, about 5 years ago. Nothing is new, so why does holding this make his absence feel that much worse?

I look to the top right of my peripheral out of habit. I look behind me out of habit.

No blinking red dot. No Grandpa.

Right after Yéye stopped diving with me, he started waiting for me on a grassy mound near the shore. He used to call it a wedding cake because of how the mound looked; concentric, stacked with three layers. But the sides weren't smooth like fondant. It was tilted, patched-worked, sun-bleached sand on one side, dry sedges on the other – the top flattened by years of him sitting on it while I swam. As he waited for me, he saw everything through the connection from my lens.

I want to see his ghost on the top tier, wearing a faded baseball cap, a toothpick in between his front teeth. Or I want it to be empty, so I could long for him. But the mound is busy. Naked mole rats nap and play in their freshly dug burrowing holes, a collared lizard sprawls on a rock. Barrels, prickly pears, tall saguaros – I don't know where they came from – they were never there before, and now they're naively happy to be in Yéye's old spot.

Which is worse, to be abandoned or to be forgotten? Or should I tell myself, Grandpa brought out life wherever he went?

On the shore, I splash some water on my face. I whisper, 'We are one. Love for the desert. Love for each other.'

The oasis tastes sweet, pure, something fruity (probably the arfajips), nothing the CondensAir™ can replace. I shake my head. If there were no CondensAir™, there would be no Country, Capitol, or me.

I bite down on my oxygen regulator and fall into the lake back first, sending a school of minnows into a frenzy.

The Capitol's skyline ripples underwater as if I'd been in an illusion after all: if I splash into the lake and no one is around to hear it, does it make a sound?

Every year, I'm a lost mermaid returning to sea.

Sunlight laces through the water, shimmering right through it like a ghost.

I'm in a dream, a quiet quiescence before waking.

Lake Qing lives up to its name, the Chinese character for clear, because hundreds of algae and fungus are at work, cleaning the water for the arfajips to receive sunlight without any obstructions.

The belt helps me sink lower, but anxiety crawls at my throat, like I'm wasting time doing nothing, so I push, swimming harder to the bottom until I finally see *them*. The arfajips are partially named after the golden flowers that bloomed in the old Kuwait desert. Two plants in unrelenting terrain. But arfajips don't have petals, they don't bloom with a promise of pollen. Instead, arfajips look, well, dead. Like their color has been sucked by the desert. They stand a foot tall, an empty stem stripped of leaves, with hundreds of tiny needles at the tip.

At the sediment floor, my fingers skim their needles; they bristle like the edges of a paintbrush. I wonder what the arfajips are saying to each other, if they're thirsty and alerting their friends that they can point their needles toward me. Arfajips share secrets through a network underneath the lake, their roots entangled with each other, holding hands, like the way I used to wrap my whole palm around Yéye's pinky.

Anxiety, or grief, I don't know what it is, but something squeezes my throat tighter.

I swim a few feet above the arfajips, and with the help of the belt, I lie face up and

fall,

fall,
fall
onto the needles.

There's a slight pressure on my back, and my body arches, shivering in a crescent moon. The stems pierce my skin, but the arfajips never hurt me. Acupuncture, Yéye used to say. The flowers were not designed to hurt their prey, but to indulge them in euphoria for their victims to come back, obviously.

The best part of watching the arfajips drink is the bubbles they form, in an extra large kind of size, like a pomegranate. The idea is simple, but delicate in its complexity – a wonder in evolution. The arfajips harvest water from me and release my water as bubbles back into the lake. The bubbles don't float towards the surface though; they are weightless, the same density as the water of Lake Qing. They crowd around me like I am a friend, or a flame to a forest of moths.

When one of these bubbles floats up to my side, its walls shimmer in a rainbow prism, my nose and eyes distorted in the reflection.

I stretch my fingers out and pop it.

Nothing visible comes out from the bubble, but I know my water is released back into the lake.

I count another ten bubbles.

I pop them all.

I count a hundred bubbles, a hundred and fifty – and then they've swallowed me. I'm taking the galaxy's largest bubble bath.

Yéye is suddenly next to me and we're cackling together, popping as many bubbles as we can. Two people's worth of water must be at least… A MILLION BUBBLES???

Pop.

Pop.

Pop.

Yéye smiles at me with his eyes. That's his trademark – his eyes gleam like the Capitol. 'Don't forget to live for me, Tian.'

Live, live, live.

'Yes, I'm living now!' I say.

We used to be afraid of the desert. We hadn't believed in desertification until deforestation and unsustainable agricultural practices led to the erosion of the world's coasts. It happened slowly, and it's funny to think how no one did anything about it. There were too many powerful people taking too long to make decisions, or making decisions that made the problem worse, and gradually, cities fell, and once cities fell, chaos led to the inevitable fall of known civilization, and then nothing. An unspoken history ripped out of a book. Centuries later, when people returned to this empty city of old Toronto, they found something that wasn't there before. An oasis with real, drinkable water. And inside the oasis was a treasure.

Sure, in the old days the anemone provided the clownfish with shelter, and the clownfish returned the anemone with food, but mutations after the desertification forced a new design, a new mutualism between multiple pairs of species. When arfajips take water from their prey, they return some of the prey's water to the lake, so the home will always be a home.

Over the years, people flocked to greet the arfajips and give water back to both the plant and the lake in reciprocity, but when the Capitol got big enough, and we could manufacture our own drinking water, the tradition stopped.

But not for us; Yéye and I completed this ritual every year.

With Grandpa, the bubbles, and my oxygen tank, who could not be perfectly happy?

I close my eyes and I hope the oasis can hear me.

Thank you for the water.

When I swim back to shore, the setting sun reflects over the lake in flames and honey, and the high is settling in.

I spit out the receptor into my hands, but it falls onto the sandy shore instead.

'Oops,' I chuckle.

I look over to the mound. No Yéye.

But there are tons of prickly pears.

'Oh!'

I climb the wedding cake. With the way the cactus' stems juts out like two arms, they look cuter than I remember them to be. I decide in the moment that I do indeed like the pink bulbs on top of their heads, so I plant wet kisses between their sharp spines and giggle and giggle until my cheeks are sore.

I bow to them. 'I'm sorry I was rude before. I love you all very much too.'

On the branch of a nearby mesquite, a monarch butterfly shakes herself out of an abandoned cocoon.

'PRETTY—' I cover my mouth.

'Pretty,' I whisper instead and cackle to myself.

I hold my index finger next to her legs. I wait, and wait, and listen to the bees' hums, and I don't know how long I wait before she finally crawls onto my finger.

Look at the pattern of her wings! How could circles exist so *round* in nature? Wasn't Opal wearing butterfly earrings?

Another red-tail hawks *keeeeears*.

In the lake, Yéye told me to live, right?

'CROW,' I finally say, 'I need your help.'

'Yes, Tian,' CROW never misses a beat.

Maybe I can take off with the monarch butterfly into the sunset too. Maybe I'll find Yéye at the horizon between the sky and the water.

'Send a message to Opal. Tell her I'll be right there.'

Kelley Tai (she/her) is a science fiction and fantasy writer who daydreams about alien plants and wishes she were part-cat. She runs her own freelance editing business at Bramble and Crow Books and is the Online Editor for Augur Magazine. When she's not busy herding her two kitties, Kelley occasionally makes videos on YouTube about the books she's writing and reading. You can find the list of her published works at www.kelleytai.com.

If Cooler Heads

Sara Kate Ellis

Our town still had an estuary when I met you, a broad stretch of brackish water reaching out into the open sea. In those days, we could still see a scattering of foxglove in the marsh grass and maybe a few cormorants perched along the walk rail as we made our way to the shore. That was years before the water took my parents' house, before the sea sifted in, submerging the inlet and turning the blue into a changeless stretch of gray.

Some of it remains. If I squint hard enough, I can fill in the emptiness between the mudstone cliffs, imagining the rows of clapboard houses that used to look out over the ocean, and the old brick schoolhouse where I met you.

You were dressed for the cold, even though our Octobers had been hotter than 70 degrees for years, wearing a wool-lined denim jacket that smelled of mud and hay. You'd come from a cooler place, but not by much. A landlocked state where the water found you all the same. It had taken both your parents and swept you up here, to an uncle who set you to work on his boats barely a week into your stay.

'She won't last a month at school,' my mother said. 'Dan Arden will make sure of that.'

Arden ran a struggling fishing charter near the boardwalk, a mad Ahab with a work ethic from another century. A few of us had worked for him in the summers, helping the dwindling tourists don their safety vests and cleaning their pitiful catches of salmon and rockfish, but most of us quit after a week or two.

You wouldn't have that choice.

Our teacher, Mr. Coaley, gave you a terse but chummy introduction, as if he was already counting the days until you dropped out. Then he left you adrift to pick a spot among the many empty spaces to choose from.

Families were migrating away from coastal areas and our numbers were dropping, dividing into two camps of privileged and desperate holdouts. I thought you'd stay up front where Mary Alice and the others traded prescription pills and texted within full view of the teachers, but instead you took your time, scanning the room as if you were looking for a place to live. Your eyes rested on me.

The others sniggered, turning in their seats to watch your progress as I locked my gaze on the whiteboard. The words were dashed across it like a threat scrawled on a bathroom wall. *The ice was here, the ice was there, the ice was all around...*

We were studying *The Rime of the Ancient Mariner*, and Coaley segued into a recent incident of an albatross harassing a boat near Lincoln City. 'Scared the hell out of the crew. They took it for a coastal UCAV run amok. Thought they'd get shot at for poaching. First albatross in these parts in over a hundred years,' he said. 'What do you think, new kid? Is it an omen?'

For a long moment, you said nothing, and I could see him taking on that smug but pained expression he got whenever he called on a slow kid. But you slung your backpack on your chair, allowing yourself to settle in before answering.

'No point.'

'In what?' Coaley snorted. He had, after all, condescended to make you feel welcome.

You nodded up at the words on the board over which someone had doodled a bird pooping in flight. 'Omens are lip service,' you said. 'Reassurance that we won't be as stupid.'

I laughed, expecting the others would too, but my voice skipped over the silence like a stone across a stagnant pond. Coaley said nothing, mistaking the quiet for vindication. He dabbed at his forehead with a handkerchief and turned back to the wall. You leaned a little into the aisle and I caught the lonely

14

slash of a rope burn on your palm as you held out your hand.

'Thanks for that,' you whispered.

I shoved down my smile, pretending that I hadn't laughed at all, but if it hadn't been for the heat, I might have recognized the warmth inside me for what it was. Instead, I gave you one of those looks, the kind that accompanied all the 'I can'ts' and 'tireds' and 'dones' and 'nopes' we found so funny at the time, offloading the words for their implications. In truth, we were all slowing down inside, the heat sapping our thoughts and who we were to each other.

By the next week, the rumors were already circulating. That you were dangerous. That your parents hadn't really drowned but were addicts or illegals who'd gotten themselves deported. We ignored and obsessed over you in equal measure, for you were beautiful, despite your oddness, with dark hair that trailed down your shoulders and a calm so many of us were ceding to crying jags and outbursts that came from nowhere.

Our school was already trying to temper our behavior, even as pundits blamed these mood swings on tech and bad parenting. The nurse led workshops to teach us breathing techniques and passed out pamphlets, advising us in veiled and ominous language that it was 'best to avoid crowds' when the temperatures were high, and that important decisions should be 'deferred until after sunset'.

'What? Are we vampires?' Mary Alice said. She was eyeing you across the room as if you were the real monster among us. You'd given up on making friends by then and sat away from me in the back, ducking out sometimes even before the bell rang. No one, not even the teachers, tried to stop you.

When we spoke again, it was mid-November, the air less humid, but never the sweater weather our parents spoke of so nostalgically. Our class took a yearly field trip to the Marine Science Center, where we prodded anemones and half listened as a volunteer lectured us about attempts to save the sea star.

We stood around a water tank, checking our phones and not

paying attention as a grad student held up a specimen, poking his fingers into its shriveled nubby limbs. We were too busy planning our escapes, to colleges that still taught person-to-person and threw their money into small self-contained cities with controlled temperatures and covered walkways that meant you never had to go off campus.

I didn't doubt I'd get in somewhere, but my options were shrinking. My father's company distributed outboards and equipment for smaller boats: Walkarounds, Caddy Cabins, and Dinghies, but that once-lucrative business was halving with each storm surge. Now when we sat on our broad porch looking out over the ocean, he seemed less like a man surveying his kingdom, than someone nervously gauging its distance from the cliffside.

I edged away from the group and strolled outside, losing myself on a side trail that wove along a granite revetment. Above it, the old coastal highway still clung to the edge like concrete bones, all but parts of it having crumbled away. Below it was a cluster of tide pools, guarded from the sun by a cave system which afforded a view of the sea. I'd been down there many times, but never alone, and as I made my way along the edge of a pond, I caught my reflection.

You were in it.

Shimmering. Your hair tied back and your eyes bright with excitement, as if you'd been trapped down there and I was the first human you'd seen in years.

I whipped around in surprise, my shoe slipping against the rock, and felt that rope burn warm and firm around my wrist as you steadied me.

'I was late for the talk,' you said, as if reading my astonishment. 'Figured there was no point in staying. Not for that.'

'Oh …' I felt my face grow hot again. 'It really isn't all that interesting. Depressing really. They tell us the same things every time. Only thing that changes is the fish get smaller.'

'Thought so. Yeah…' You let out a nervous laugh and nodded up at the ridge of broken highway. 'But there is something else I'm curious about.'

You smiled and I smiled back, and this time it stuck. I don't know who moved first, but we both turned back, slipping through a tear in the fence to climb an overgrown rope trail winding up the cliffside. At the top, you hoisted me onto the concrete plateau, a chunk of weed-dotted asphalt shoring up the remains of an abandoned service station.

'These used to be all over the place.' I ran my hand over a rusted-out gas pump, 'So many of them along the highway. Just standing stones now.'

'Vestigial?' You stretched your lips back, doing a perfect imitation of the volunteer at the Center. He'd repeated the word incessantly, nearly spitting on us as he held up a scrawny specimen of Pacific salmon, pointing out the adipose fin near the tail.

'So you *were* there.'

You grinned. 'Only for a minute.'

I wanted to laugh, but my mind was swimming with the way you looked, the gentle set of your features as your gaze fixed on something in the distance.

Even from that far, they were enormous. Two albatrosses. Solitary on the hunt, but flying close together in what seemed to be coordinated movements.

'Maybe they are coming back,' you said.

When I was small, my mother made me take calligraphy lessons. I never got much better at penmanship, but I discovered a cold truth. Those lines of ink weren't a continuity, but separate strokes forming the illusion of a whole. I remembered that as you took my hand, our fingers interlacing like the shadows of those birds as they swooped low over the waves.

After that day, I started to notice the ruins in our town: the double staircase on the beach that wound around its out-of-use counterpart like the mortified half of an alcoholic marriage, the phone booths, wires exposed and scratched with graffiti, and the benches still marking long defunct bus routes to towns wilted from existence. It's true that places shrink as you grow into them, but this was a shedding, the chunks that once served intelligible functions breaking off like the ice floes now reinventing themselves as hurricanes and erosion.

I looked for you in school the next Monday, but you were gone days, then weeks, and finally a whole month passed. When you came back, I had a boyfriend.

Nick was a test run, one of those jovial, barrel-chested types who liked to roughhouse with his buddies and tongue-kiss me in front of his friends. All of us had them: girlfriends, themfriends, and boyfriends, hanging about our necks like the bird in Coleridge's poem.

'Sex should be weighty,' our health teacher told us. 'It's an expression of love.'

But if I felt unimpressed, the others talked about their romances as if they were chores, detached from their pleasurable purpose in the way that frosted shop windows at Christmas harkened back to snows we couldn't remember.

I wondered, even hoped, you might be jealous when you saw us, like the albatrosses I'd read about, forced to fly out further and further to find food. Returning home to find strangers guarding their nests.

They used to mate for life.

But you were just gone again, an absence that coated my insides with a listlessness, like the humidity that seemed to drain our days of purpose.

I thought Nick would be okay with my ending things, but like a crab whose shell grows fragile from the surf-borne acidity, he broke down, melting into a sweaty, sobbing bundle before my eyes.

'You can't be saying this,' he said. 'You – you didn't let on at all.'

I glanced around the pier, embarrassed and berating myself for not having planned better. I'd followed all the advice in the *Weathering the Storm* pamphlets our school passed out, hydrating to avoid bad decision-making and venting in my journal first to filter out the ugliest bits of what I wanted to say. But Nick was pleading with me, his fingers tight around my arm as I tried to back away.

'You didn't let on at all,' he said again, as if that gave him rights. As if I'd forfeited my right to say no. I didn't hear you coming, but there you were. Standing on the boardwalk, looking scared. You held a fillet knife in your hand like some Final Girl in a horror franchise.

I screamed and you dropped it, shoving your way between us. Nick sprang back, less from the force than the stench of you. You wore an apron slathered with fish guts, and the ammonia made our eyes water. He raised his hands, a self-righteous glower on his face as if all those rumors about you were true.

'Okay,' he said. 'Okay.' He turned to me, expecting to find reconciliation in our mutual disgust, but I held a hand to my mouth, and forced myself to breathe.

'Go.'

He did and you waited, your body coiled until he was out of sight and another stretch of silence threatened to fall between us.

'I'm sorry,' you said. 'I didn't mean to scare you.'

'You didn't.' I nodded toward the knife on the pavement. It had slid across the water-slick boardwalk, stopping at the edge of a drain. 'Startled mostly.'

You laughed, relief washing over you, and hurried over to retrieve it, giving it a quick wipe with the edge of your apron before you tucked it into your belt. 'Do you need anything? A bathroom? A glass of water?'

I shook my head, and maybe it was the excitement or just the sight of you there, but my pulse caught up with me, sending the boardwalk into a spin. I listed and you stepped in, keeping enough distance to avoid your apron touching me, your fingers soft and steady at the small of my back.

They stayed there as you led me through the back entrance into the cleaning room of your uncle's business, where the flash freezer hummed and the lights cast a somber pool over a stainless-steel cleaning table.

You topped off a cup with crushed ice, shaking some of it into a cloth. You tried to press it into my hand first. You didn't want to presume, but the heat makes us rash. It ruins our ability to think.

Know your triggers, our teachers warned us. *Stop and play it out in your head before you act.*

If it had been cooler that day, a normal temperature from, say, the 20th century, I wouldn't have done what I did. I would have talked to you first. I would have learned more about you, so I'd have more to remember. I wouldn't have taken your hand and pressed it to my cheek, or leaned in, watching your dark eyes grow large with surprise as I kissed you.

When you pulled away, I wouldn't have shoved you off as if the whole thing was your fault before walking away.

The weather works its way inside you. It changes who you are. I can see that now. Each time I come here, I want to tell you.

That wasn't me.

That wasn't how we were meant to be.

People still insisted that we never got hurricanes. They clung to that falsehood even as the winds toppled their cars and their porch swings flew through their windows like flails. In May, our teachers herded us all into the basement, a dank concrete cavern where they set up a buffet table and wired up a makeshift sound system to play pop hits over the storm. A single laptop was open on a folding table, ready to display the destruction outside as if it was happening to another part of the country.

We threw our backpacks and coats onto the floor, assembling our cliques in a near identical layout to the spaces we took up in the cafeteria above us. The underclassmen were giddy, shrieking every time the lights flickered, but we had more serious things to discuss, like our college choices, now whittled down to a few, and who we were bringing to the Spring Formal.

Mary Alice was going East. A fire had devastated her first pick, tearing through the campus, and destroying most of the housing. We'd watched the feeds of students fleeing on the heels of bobcats and coyotes, of the smoldering library building whose narrow windows – built to fend off protesters – made the remains look like a charcoal grill turned on its side.

'Decided yet?' Mary Alice asked me.

I shook my head, feigning distraction. Across the basement, a group of kids was playing a fortune-telling game with a deck of cards – *who will I fall in love with? Who will leave me?* That's when I saw you.

You were leaning against the wall at the far end of the room, looking for all the world like a ghost. Mary Alice drew up, craning her neck to see where I was looking, but the lights flickered again and went out and stayed that way.

We fell silent as the music juddered to a stop, until one kid screamed, and another one told him to shut up, and soon both were drowned out by the storm as it rattled the egress windows. A flashlight and then several phone lights flickered on, like a network of cities as seen from space. You were still there, still real.

I scrambled to my feet, hands out as I cleared Mary Alice and the others and the pile of backpacks amassed around us like sandbags, feeling my way along the surface of a table as my eyes adjusted to the darkness.

If cooler heads had prevailed, I would have maybe just held your hand until the storm died down. But in that instant, the wind sent a branch through one of the panes, filling in the places where our voices failed us, and I pulled you to me and kissed you.

If I'd had my wits, I wouldn't have let you lead me into the space behind the boiler, or let your hands travel up beneath my shirt, your fingers cool against my rib cage. I wouldn't have pressed closer to you, drinking in that scent of hay and football leather that clung to you like your past.

I wouldn't have stayed there, hiding away from the others until the all-clear sounded, or let you lead me through the debris-strewn street, past the jam of worried parents and emergency vehicles to the pier and that room, lonely and spartan, above your uncle's shop.

The sound of sirens became our sunlight, fading as if to signal night. There was no winter to contrast summer anymore, no chill to set the burn in my chest into relief. There was just need and ache, and feeling that something had been taken from us.

You trailed that rope burn over my body, your lips soft and warm against my skin as we lay together on your narrow bed. I remember thinking that I'd never shivered like that before. When I awoke, you were leaning over me, a smile playing at your lips before you spoke.

'I'm going away for a while.'

'Where?'

It seemed like such a strange thing to say. You'd been going away for so long.

Your hand rested on my shoulder, generating a noticeable warmth that long ago might have been a comfort. 'My uncle's about to lose the shop. And there are better hauls out in the No Zones. Bluefin and halibut. Enough to pay off what we owe and then some. There's a boat that needs crew, people who aren't afraid of the patrols.'

I froze and you smiled, a little happiness breaking through at my response. That there *was* a response. 'That's why I came to school. I wanted to see you.'

If the weather had been colder, I like to think some determination might have seeded its way into me. I might have been brave enough to tell you that you were as close to perfect as I was ever going to get.

When we watch a clock, we've got the solace of the hands breaking time into manageable segments. Before that, people measured the hours through shifts in the light, or navigated by the salt scents drifting off the ocean. But when everything is changeless, we are lost and flatlining.

Nothing but a single letter from you ever arrived. The old-fashioned kind, written in a hand so smooth, I questioned if it was yours.

Poachers cut off electronic communications to thwart the Automatic Identification System and evade detection. The authorities were efficient, if not brutal, with transgressors.

You said you'd had a good start, and that if things kept up, you'd be back within a month. You said you loved me.

I wrote back every day, answering in kind, leaving your letters with Arden, who seemed baffled that you even had a friend. He promised me they'd get to you. I promised myself that I'd wait for as long as I could.

In June, a row of houses slid into the sea and my parents sold before the property value dipped further. I cried over it, but they just said I was going to college, as if that somehow disqualified me from caring.

I stayed on in the empty house for a few weeks after they moved, scrolling the dial on Marine Radio and listening for news of ships in trouble. In that last week, someone smashed the windows of Arden's shopfront. He boarded them up and didn't bother ordering replacement glass.

He would tell me later it was the birds that finished you. The albatross wasn't making a miraculous return, but being bred and trained to track ships who'd switched off their AIS. They were fitted with lightweight radar, and drawn to the smell of the catch, they'd ping the poachers' locations back to the authorities. The hunter drones did the rest.

You poached. You sank.

I wonder if when you saw them, you remembered what you'd said about omens that first day in class. I wonder if you thought about our fingers twined together as we watched their crisscrossing paths form the mirage of something whole. I hope you knew I loved you, that I still love you, even if you never got my letters, even if love, like everything else, is vestigial. An abandoned bus stop or an empty phone booth, some remnant of the old shoreline where you hold a shell to your ear as you gaze out at that endless, flat expanse.

Sara Kate Ellis is a Lambda Emerging Writers Fellow and attended the Milford Science Fiction Workshop in 2017 and 2022. Her recent stories have appeared in Analog, Fusion Fragment, and Metaphorosis. She teaches at Meiji University and lives in Tokyo with her partner and two ornery street cats.

Pure White, Ocean Glass

Tania Chen

April's pale moon arrived with a yawn on the night of their birth. Tiny hooves pressed against their mother's shell mouth, prying it open from inside as they blinked their blind eyes open underwater. Umbilical cords connecting them to the fleshy innards pulsated, dissipating. The unicorn kicked away from their mother and sank to the bottom.

Sand and silt lifted as they darted along; four slim legs and a tail that seemed seafoam but was all sharp edges. Curtains of sea snakes making way as their bursts of energy changed the tides.

Their smooth, shell-white coat reflected the purple of moonlight through water; over the seasons it took that colour, kept it – a slow dye.

They were caught and hauled onto a tiny wooden boat before being unceremoniously placed inside a glass bottle and sold. A label in conch green: WATER-PURIFYING-UNICORN.

They were given to a little girl who tapped the bottom, sliding them into the pond water – no longer salty as they were used to.

The small house's pond was designed for carp, for a red-slider turtle or two. Not for their kind.

'Oh, thank you, Papa, Santiago is pretty.'

'Santiago?'

'Yes, don't you think it looks like a Santiago?'

They thought the name to be fine; acceptable for the moment.

Living in a pond stunts Santiago's growth. They are fed little white grains of rice as they splash around the surface, horn spearing through anything that got near. Teenage rebellion or just incandescent rage.

But true to the bottle's label, the waters of Santiago's pond never turn murky, the fish never get sick but die of old age and all the plants – within and without – are a lush green, thick with flowers.

Santiago arched their neck and nipped off an orchid newly bloomed.

The little girl, her face now adolescent giggled. 'You're going to ruin your appetite like that.'

Santiago knew that, but the flowers were so sweet, unlike the kelp deep along the coldest depths, and dim compared to the sweetly bright coral reefs from home. They missed it.

'Don't you feel a little sorry for them?' Next to the little girl's face was another, round and freckled. Santiago paid them no mind, there were more flowers to chew through after all.

Now an adult, the little girl came less to the pond. Her father had passed and left the property to her, and she in turn had finally married her heart's desire: the other little-girl-now-adult who lived down the block and snuck teaspoons of sugar for Santiago. They liked her for that, and now that they were both living under the same roof she came often to take care of Santiago.

It's dull, sometimes, confined to this space when the whole ocean should be their home.

'I'm sorry you're here, I tried to talk Miranda into letting you go back but... you're the last gift she has from her papa.'

Santiago did not understand what that had to do with freedom, turning inkwell eyes in her direction: impatient, reproachful.

'Saw the news the other day, things are getting real bad in the sea without you guys taking care of it.'

A snort, followed by an angry kick of their hind-legs, splashing water on her.

Santiago learned her name after a few visits: Lenora. That was the first human name they cared to remember.

Lenora came to feed them, tell them the news of the world outside the pond and garden, how everything rotted as their kind were put in bottles and sold to private collectors.

Miranda and Lenora often fought about how selfish it was to keep a unicorn in the pond when their kind were meant to purify oceans and seas and lakes and glaciers – endless bodies of water that transformed with the passing light of the moon.

One night, Santiago felt a splash in the pond, the smell of rubber boots and the unceremonious attempt to shove their body into a flower vase: narrow neck and oval bottom. They kicked, crocodilian hissing, and it was just as disorienting as the first time. The past flashing in their memory; from sea to bottle to market to pond.

Santiago snapped their jaw closed, tasting skin and blood.

'Oh shit, Santiago, ow—' Lenora dumped a bougainvillea bud in as a peace offering.

Santiago settled, munching it violently between flat, grinding teeth.

Lenora reached for Santiago once more, hands cupping the tiny milk-white body.

'Sometimes,' she begins with gravitas in her voice, slipping Santiago into the vase, 'not letting certain things go isn't the right choice.'

Santiago's first and last car ride brought a change in the air. From the smells of the pond and city, to the familiar, scent of salt. They knew where they were going: home.

From pond to vase to car to sea. April's pink moon casting light on their path as Lenora stepped onto the beach. 'I know Miranda is never going to forgive me for this but...' she turned away, hiding tears that Santiago could taste in the air.

All the other people on the beach with their bottles and their contents: tiny scrabbling white bodies yearning to get back to the ocean. Their chest expands, horn tapping impatiently on glass, and their thoughts rise like warm air:

Beautiful how revolution appears.

Tania Chen is a Chinese-Mexican queer writer and Clarion West Workshop 2023 graduate. Their work was selected for Brave New Weird Anthology by Tenebrous Press, and has appeared in various places like Apparition Lit, Strange Horizons, Pleiades Magazine, Baffling Magazine. Currently, they are assistant editor at Uncanny Magazine.

Once Upon a Biofuture: Tales for a New Millennium is an anthology of stories, from fiction to memoir, by a multidisciplinary team of scientists in the UK Centre for Mammalian Biology at The University of Edinburgh

A mix of biology, philosophy and mythology, they explore a powerful new technology that is re-engineering our world: synthetic biology. Many of the stories are biographical, offering insight into how the scientists become scientists and the lessons learned through science exploration, or taking us to new imaginative worlds

The stories were recorded and transcribed, or workshopped and edited by Jessica Fox, former storyteller for NASA, and artist-in-residence at the Centre. This unique role enabled the scientists to lead the storytelling and retain creative control while being guided by an experienced writer.

Once Upon a Biofuture: Tales for a new millennium

Edited by Jessica Fox

Published by The New Curiosity Shop/ Shoreline of Infinity Publications

ISBN: 978-1-7396736-6-6

RRP: £12.00

A Change of Direction

Rhiannon A Grist

Eryn gave the grub's hub screen a whack with a mittened hand. The swirling progress icon did cheery laps around a flickering, pixelated smiley face.

'Stupid piece of shit.'

She'd told her boss to go with a Nordic replacement for the broken interface, but they'd fought her down on cost. Eryn had wanted to argue that it was this kind of cheap, myopic thinking that got them into this frozen hellscape in the first place, but she was long done fighting. She just drove the grubs now.

The defroster wiring made good work of the ice on the windscreen, just in time for Eryn to spot her new apprentice shivering at the entrance to the garage. Eryn winced at the skinny wee thing. Kids were short these days. Used to be they'd tower over the adults by the time they reached fifteen. Now most twenty-year olds barely reached Eryn's broad shoulders. Maybe with the improved crop diversity, the next generation wouldn't suffer quite so much.

Eryn flashed the lights of the grub. The kid startled then scurried toward her, padded coat flapping uselessly around her legs.

Alright, Eryn thought, *first lesson: proper wrapping up*. Not the most interesting element of energy reclamation, but it was the difference between getting home with a touch of frostbite and not getting home at all.

The kid wrenched open the door, using half her body weight to do so. Eryn's dodgy elbow ached just watching. A cold blast jumped up into the seat with the kid, and the grub complained about the broken heat seal. Eryn muted the alarm with a tap as the apprentice pushed a full-to-bursting rucksack under her seat. The kid pulled the door closed then took in the crowded cockpit of the grub: a hodgepodge of soldered together panels, heating pipes and storage crates. It probably smelled fusty as all hell, but she'd get used to it. They all did.

The kid looked Eryn's way. Another fresh face. Gaunter again than last year, but the eyes were just as bright as all her predecessors.

Eryn nodded to herself. *Right then, here we go again.*

'Hello, my name's Eryn. I'm your trainer.'

She remembered to smile. There would be time for learning the turbine engineers' code of grunts and curt nods later. The aim for now was not to scare her off at the first meeting. They needed more energy reclaimers than they could grow.

'Hey, uh, hello Eryn.' The kid's speech whistled slightly through a gap between her front teeth. 'I'm Cooper. But you probably knew that already.'

Eryn did in fact know that already. She hadn't quite got to the point of not bothering to learn her apprentices' names, though she had skimmed pretty much everything else.

'Cooper. Don't hear a name like that often.'

'Yeah, my mum hoped I'd get to work with her on the chicken farm,' she grinned.

Eryn hadn't heard that one before. New adults were assigned

job roles based on their skills in school, not their name. They had some level of choice, a range of roles suitable for their abilities plus whatever the hab needed more of. But it was nowhere near the dizzying array of choice Eryn had had when she left school. Perhaps Cooper's mother had Cooper during the early days of the new role allocation system. It had been, admittedly, a little vague to stop the community crying fascism.

Eryn nudged her shoulder. 'It's ok. I know this isn't everyone's first choice.'

'No, I want to be a turbine engineer! I worked really hard in maths and science and everything.'

Eryn squinted at the kid. 'You want to work in Energy Reclamation?'

Cooper nodded brightly. 'Yes.'

Well, isn't that interesting.

Eryn set the grub into drive and pulled up to the fuel-permissions window. She flashed her pass and the work authorisation on her slate. The bored face on the other side nodded and pointed her over to pump two. Eryn drove over to the heavily guarded middle section of the garage. The gate lifted and the pump tech behind it waved her over. Cooper leaned against the window as the tech lifted the grub's fuel cap and connected the hose.

'What are we doing?'

'Fuelling up before we head out.'

The irony wasn't lost on Eryn. They were living the consequences of humanity's over-reliance on fossil fuels. Now here she was, getting those same fossil fuels pumped into the tank of the grub. Everything in the habs worked on electricity. That was good enough for lights and hydroponics and heat pumps, but electric buggies could only last an hour or two out in the cold.

For chewing through miles of ice, they needed diesel.

The pump shut off at the end of her allocation and the pump tech waved her out.

Eryn understood the necessity of it, was careful with her allowances, always abided by the strict fuel waste regulations. But she always felt a little bit dirty every time they filled the tank. She wondered what the kid made of it. Would she see judgement, blame, on Cooper's face?

Cooper was sat cross-legged in the passenger seat, busily scribbling in a small notebook in tiny handwriting. Eryn huffed and noted her boots were already off. She made a mental note to add boots to her "wrapping up" lecture later on.

Eryn marked the main journey targets in chalk pen along the rim of the fuel gauge. Then she buckled her harness and tightened the kid's.

'Alright. Let's go.'

The garage door opened out onto a dark cavern of ice with walls of compacted snow reinforced with orange-painted scaffolding. A couple of lights flickered on, illuminating the space outside the arc of the grub's lamps, but the tunnels beyond were pitch black and deadly cold.

Eryn flashed the grub lights and waited for a signal from the north-bound tunnel.

None. It was clear.

The grub rumbled, tyres crunching over new frost, as Eryn drove into tunnel. The only illumination came from the grub's headlights. All that could be seen in that small pool of light was twenty feet, forward and back, of pale compacted ice and frozen tunnel walls. Eryn could almost feel the silence as the cold closed in around them.

Then she heard it.

Cooper was humming.

Eryn gave the kid a sidelong glance. She hadn't had a hummer yet. She'd had an anxious nail picker, a compulsive liar. One apprentice had sat completely still and silent the whole time, eyes staring forward, unblinking. But Cooper was humming, tapping her pencil along a page of her notebook, engrossed, unphased by the terrifying world outside her window.

Eryn followed the tunnel to the north-west electrical substation. The lights came on bright and stark as they approached. Cooper slapped her hands over her eyes.

'Augh!'

'Sorry!' Eryn winced. 'Shoulda warned you.'

She always forgot to warn them.

The substation was covered by one of the same strong polymer domes that protected the habs. Eryn had been on the crew that had painstakingly tunnelled the route to this substation. They'd lost two people putting up the dome.

Eryn nudged Cooper's elbow. 'Time to wrap up.'

Eryn showed her the proper way to prep for the cold outside. Tucking the inner thermal layers into each other. Then the middle layer, thick and fuzzy for warmth. Then the outer layer, tough and waterproof with tight seals around the double-layered gloves and lined boots.

Sufficiently bundled, Eryn popped out of the grub with the keys. Cooper toddled out after her, bouncing along, unused to the added padding. It seemed bizarre to Eryn that they still locked it up. As if anyone without a grub would survive the journey.

The gate swung open and Eryn led Cooper inside. They navigated the tight corridors between circuit breakers and transformers, to the right incoming submission line. Eryn pointed out the buzzing lines overhead.

'Make sure you don't touch those. They're a helluva thing to fix.'

Cooper looked up warily. 'Do they always hum like that?'

'When they're working,' said Eryn. 'Bit like you.'

Cooper smiled a little but stayed hunched. Eryn thought about how strange it was, relying on something so dangerous. Something that could kill them in an instant. Eryn wondered if life had ever been different, if existing had always been this uncomfortable balancing of power.

She checked the connection to the proposed farm for turning. A trickle of current was still making its way through; a small kinetic charge from the blades swinging in the wind.

'That's the one,' said Eryn.

She confirmed the location again on her slate, double-checking against the coordinates she'd been given in the garage. There it was on the latest satellite images, a little beacon of hope.

Thankfully, the satellites were still working. Who knew what they were going to do once their orbits decayed. They still had the knowledge, but would they have the skills, the people with real-world experience, to build another satellite? More importantly, would they have enough fuel to send anything up?

Eryn sighed. There would be no more satellites.

Cooper started humming along with the power lines above.

Eryn straightened up. 'Come on. We've got a long way to go yet.'

The tunnels got rougher the further they drove, until Eryn reached the point where the sat nav pointed them off the beaten path. She fitted her ear defenders, then directed Cooper to do the same.

'Alright, kid. Watch this.'

With the latest satellite image loaded up on the hub screen, Eryn went through the controls to unpack the drill from its tidy standby position at the front of the grub. It unfolded into a whirlpool of metal teeth. Eryn primed the drill and slowly raised the speed, warming up the rotating parts, the jagged teeth turning in opposite directions. She gently drove the grub forward, edging it into the wall of ice. The bit made contact. A high screech shook the whole grub, as a shower of ice chips sprayed up over the windscreen.

Cooper's eyes widened. Most apprentices had some reaction to the sound. Cooper and the rest of the kids who'd grown up in the habs were used to tight communal living. That meant watching how much space you took up, shrinking your gestures, and keeping the noise down. But drilling through ice was big and noisy. The earlier apprentices had rarely been bothered by the noise, holding some ghost of the bigger, brasher world they'd lost in their memory. The later ones had been so shocked by the sound they'd gone into a panic. Now everyone was briefed before they ventured out.

Eryn remembered screaming at the top of her lungs at a gig. She remembered throwing herself into a pile of writhing festival-goers on the dancefloor, arms flailing, bodies bashing into each other. She'd always been big, but at gigs she was even bigger, her mass, her energy, her voice and spirit raised to a roar in a roomful of noise. She'd been praised for it. Respected for it. Watching the slimmer guys dart out of her way as she threw herself into the fray, horns raised, tongue stretched out of her mouth as far as it would go.

She hadn't been to a gig like that in years. Decades. Maybe she'd never see another gig again. But doesn't everyone at some point go to their last gig, not knowing it's their last? She wished she'd appreciated the last one a little bit more.

Movement broke Eryn out of her memories. Cooper's head bounced along to the steady rhythm of the drill, lips pressed tight together. Eryn shifted the edge of her headphones. Cooper was humming as loud as she could, the notes just getting above the thunderous drone.

Eryn raised an eyebrow. Then she started to drum out a beat on the steering wheel. Cooper looked at her, then at her tapping fingertips, and smiled.

After a few miles, the grub broke through the permafrost and light filled the cockpit. The grub's wipers cleared the grey slush from the windscreen, revealing an undulating white desert as far as the eye could see. If Eryn didn't know better, she'd think she was somewhere in the arctic circle, polar bears and seals not far away.

That same circle was now five times larger than it was before. Before the ice cap melt hit critical mass. Before the jet stream broke down. Before the new ice age was triggered. Before the world was split into frozen wastelands, arid deserts, and one small band of highly-pleasant and highly-policed temperate land Eryn would never see.

Below all the snow were the ruins of countless towns. Hospitals, dentist surgeries, petrol stations, supermarkets, schools, houses, gardens, gyms, cinemas, food trucks, swings and slides, forests and beaches, rivers and roads, bicycles and dinghies. All lost beneath the ice.

On a far line of hills, a small sea of broken lines lay higgledy piggledy on the snow. Eryn's heart sank. That was another windfarm gone. Another safety net gone. Luckily it wasn't the one she was aiming for this trip.

'Hey, Cooper, can you mark that hill on the map?' She turned toward her apprentice. 'Cooper?'

Cooper had her nose pressed up against the glass, eyes wide with wonder. Most folk barely left the habs these days. The world outside the insulated domes was known only through photos, web cams and satellite images. It was a whole other thing to see it for yourself. As she watched Cooper's face, Eryn began to see it through the kid's eyes.

Drifts rose and fell like a lover's stomach, as diamond-white powder skittered over the bluffs, sparkling and soft. The sun came ringed in crystalline haloes from all the ice in the air. The world was white and wide. A frozen ocean of snow, all under a great sphere of sky so blue and so deep Eryn almost believed she could jump into it for a swim.

'It's beautiful,' Cooper murmured.

They reached the wind farm a few hours before sundown.

They stopped at a vast hillside, dotted with graceful white turbines, long-necked and docile like a flock of swans in a painting. Some of the blades swung a little in the breeze, accounting for the small charge still making its way to the substation. Eryn checked the clock and the windspeed. They had time to turn one.

They wrapped back up and put on goggles and pulled the muffler up over their noses and mouths. The temperature wasn't too bad just now, but a polar gust could shock the lungs.

Outside it was a clean, dry kind of cold. The sort that made Eryn's skin feel tighter, her eyesight, clearer. They surveyed the wind turbine, its blades boasting a fringe of twinkling icicles.

The better wind turbines had a yaw drive: a motor that turned the head in the direction of the wind. However, some yaw drives, like these ones, only turned 60 degrees. That had been fine before the jet stream broke down and the prevailing wind direction changed.

That's what you get for calling it a "prevailing" wind, Eryn thought.

'The generator is up in the head,' Eryn pointed up to the bulbous back-end to the listless blades.

She set out a compass on the snow. Then she took a spray can from the back of the grub and sprayed a neon pink line in the direction pointed by the compass.

'That's the way it needs to point,' she said.

Cooper looked up. 'How are we going to turn them?'

'One turbine at a time.' Eryn grinned then pointed. 'Everything in the shaft is just cables. All we've got to do is cut the shaft below the yaw head, turn it to face the new prevailing wind, then reattach it. Go check windspeed on the grub's anemometer.'

Cooper nodded under her many layers of hood and hat and goggles and muffler, then ran back to the grub to read the anemometer.

'Eight miles per hour.'

'That's fine. It's wiser not to operate above twenty miles per hour. The gusts can take you by surprise.'

Eryn engaged the grub's counterweight, a great heavy braced panel that swung out on the non-operational side and dug down into the snow. She showed Cooper how to check the brace and lock the weight in place. Eryn started up the grub's clawed crane arm. She showed Cooper how to clamp the bracing claw to the lower part of the shaft, then how to raise the larger, more complicated, cutting claw to the base of the head and clamp it tightly.

Cooper had a go on the joystick, slow at first but gaining a bit of confidence toward the end. Eryn didn't even need to adjust the positioning.

A natural, she thought with a smile.

Eryn took the controls to carefully engage the saw – a glorified tin-opener set into the cutting claw's mechanism. With a whine, it ran along the outside of the shaft smooth and careful, cutting through ice and metal.

Once that was done, she separated the two parts and checked the cut was clean with the fibre optic camera. No cables had been nicked, so Eryn engaged the turning motor and slowly turned the head. The blades juddered with the movement, large wings frozen mid-flight still trying to take off into the blue sky. Once it lined up with the line on the ground, Eryn reversed the separator and the two parts of the shaft came back together.

The sun was sinking closer to the horizon, dyeing the blades and the shaft a beautiful shade of peach. The kid was right. It was beautiful. Perhaps she'd let Cooper have a go with the cutting-claw tomorrow.

'Right, fetch the lattice,' Eryn pointed Cooper to the back of the grub. 'We'll get her patched up then stop for the day.'

Cooper scampered through the snow excitedly. Eryn looked over the field of wind turbines, shadows lengthening across the snow.

One down, she thought, *twenty-nine to go.*

As the sun went down, the winds picked up. Eryn had debated whether they should shelter back under the ice. Despite the high whistle from the grub's carapace, it wasn't forecast to be too bad overnight so they stayed above the frost.

Cooper made up the rations, clumsily spooning green, high-calorie powder into their canisters.

'Oh blast.'

She spilled some on the counter-top-slash-desk-slash-table in the cramped back-end of the grub. Eryn grunted and swept the spill into her cannister.

The tool boards on either side flipped down to create two bunks running along the length of the grub. The engine, now idle, had been modified for the permafrost life, storing the residual heat in a network of fluid-filled tubes running along the grub

bed. The tubes clustered around a small slot the size of Eryn and Cooper's food canisters.

Cooper mixed a little snowmelt into the cannisters until the paste turned gloopy, then popped them into the heat slot. She inspected the tubes circling the slot.

'That is so cool.'

'One of the upsides of petrochemical engines. They kick out a lot of heat.'

Cooper sat back up onto her bunk. She thought carefully for a moment, then asked, 'What are we going to do when we run out?'

'Of diesel?'

Cooper nodded. 'The grubs wouldn't work anymore. How will we turn the turbines?'

Eryn sighed deeply. Petrol in a cannister only lasts three to six months before it goes off. Crude oil lasts about fifty-three years. It had been forty years since the last rig fell.

'We work as hard as we can to turn as many turbines as we can before we run out. Then hope we've bought ourselves enough power, enough time, to figure out the rest.'

A jet of steam whistled out of the heat slot.

'The turbines would still need maintenance, though. We can't just leave them and hope they last.' Cooper bit her thumbnail in thought. 'Maybe we could lay tracks with charging points for the buggies.'

'That's a lot of track,' said Eryn. 'It would take a long time.'

Cooper grinned. 'I've got time.'

Eryn nodded at the air.

'Y'know,' Eryn smiled, 'it's good to see someone so passionate about this kind of work.'

'The work's alright, I guess,' Cooper fiddled with her canister, 'but that renume rate is something!'

Eryn felt the sides of her mouth slowly falling. Cooper continued.

'This is one of the few allocations that pays above basic even at apprentice level.' She beamed. 'Right now, just sitting here, I'm making much more than Mum at the chicken farm!'

All the energy and buoyancy Eryn had felt the last few hours deflated. She pushed her dinner about her cannister, wiped her mouth, then tossed the contents in the recycling port.

Cooper jumped at the clatter.

'Aw shoot, did I make it wrong?' Cooper scrabbled her spork in her meal, checking the consistency. 'I get so caught up I forget what I'm doing.'

'Yeah,' Eryn sniffed. 'Don't we all. I'm turning in. Night.'

She lay on her cot, turned out her light and rolled to face the wall of the grub.

'Oh, ok,' said Cooper.

There was silence for a moment, then the sound of Cooper quietly finishing her dinner and popping her and Eryn's canisters in the wash. Eryn watched the kid's shadow on the wall of the grub shuffle back to her bunk and scribble in that notebook of hers, until Eryn closed her eyes and forced herself into sleep.

The next morning the grub felt more cramped than ever. Even Cooper's humming eventually went quiet. The wind still whistled around the grub, not as violently as the night before but still strong enough to cause concern.

Eryn checked the readout from the anemometer several times, but each reading was different from the last. One stronger, one slower, one middling. She wiped her face and sighed.

'We can wait for the wind to die down,' said Cooper. 'I've got a lot to read up on anyway. I don't mind.'

Of course, you don't mind, Eryn thought. *You have all the time in the world, don't you?*

Eryn checked the fuel gauge. Each day spent out on the cold without a turbine turned was another ration of fuel wasted. Another ration of fuel they'd never get back.

She turned away from the anemometer and leant on the evidence of her eyes. The blades swung in the wind and the top coating of snow skittered close across the ground. A skittering. Not a stream.

'We're going out.'

'Really?' Cooper blinked at the white world outside. 'Are you sure?'

Eryn bristled at the unease tangled in her guts. 'Hey, who's the trainer here?'

Cooper looked at the floor of the grub. 'Sorry.'

Eryn had the turbine in the tight grip of the crane arm before Cooper had a chance to get wrapped up. The grub shook lightly with each buffet of wind.

'Do you want me to try the cutter today?' Cooper shouted over the rush of air.

'Nope, I got it.'

In fact, Eryn was already halfway through the cut. The base windspeed wasn't too bad, but the way the odd gust tugged at the blades had Eryn worried. The sooner this was done the sooner the turbine would be stable again. And that needed an experienced hand.

Carefully, quickly, she started the rotation of the turbine head.

'Come on,' Eryn muttered up at the turbine, as if coaxing a fledgling. 'Hold in there.'

Eryn's hands sweated inside her gloves. Her fist melded into the shape of the joystick, so careful were her movements, precise and smooth, like the gentle tilt of a bird's wings in flight. It was the best work she'd ever done.

The turbine's head juddered as it tilted to look north.

Almost as soon as it faced into the wind, a fresh gust came screaming out over the plain. The blades caught the wind and suddenly and quickly started to turn. They knocked against the crane arm, letting out a series of ringing clangs.

Eryn saw Cooper back away in her peripheral vision.

Eryn sped up the turn. The crane arm jolted. With a loud snap, a crack sprung down from the cut.

'Eryn?'

Eryn ignored her, pressing harder on the controls as if that would help the crane arm fight the growing force pushing down from the north.

Come on, come on.

She grit her teeth. She could still fix this.

The gust became a gale, kicking shards of ice up from the ground and covering the plain in an impenetrable haze. The wind whistled sharply through the rotary blades. The crack grew up the side of the shaft.

No no no no no no.

The shaft splintered. Large shards of metal sprung out from the mast with the sound of a cracking whip and the whole thing began to bend with the wind. The crane arm of the grub whined under the strain.

'Hit the release!' Eryn yelled over the howling gale.

Cooper flailed around the controls, 'Which one is it?'

'The release!'

'I can't see!'

The crane arm bent back. The grub fell onto its side. The blades came crashing down.

Cooper froze.

Eryn dropped the joystick, ran to the kid and threw them both to the ground.

The turbine fell into the snow, shattering blades and pieces of mast.

Eryn waited for the sounds of crashing to stop, then looked over her shoulder.

Almost as soon as everything had fallen, the wind stopped.

Eryn clambered to her feet and surveyed the damage, but she didn't need to look long. The turbine was lost. The crane arm was broken. They'd need to return to the garage to get it fixed. The fuel that had been carefully calculated for a full wind farm adjustment was wasted.

Cooper looked from the wind turbine to the grub to Eryn.

'Is this going to come out of our renumes?'

Eryn kicked the snow and screamed.

Once she'd been silent long enough, Cooper asked, 'Are – are you alright?'

Eryn panted, leaning on her knees, cold air burning her throat.

'No Cooper, I'm not alright.' She caught her breath. 'Each generation comes around and God help me I think, "Finally, we got it. We're going to do better."' She straightened up. 'And then you go make the exact same mistakes as all the others before you.'

Cooper looked at her hands. 'I'm trying to learn. The controls—'

Eryn swung round. 'I'm not talking about the broken turbine. I'm talking about the money.'

Cooper's eyes darted about as if trying to remember what she'd said.

Eryn pointed a finger.

'You said you chose this posting because of the renumeration. That's how we got here by the way. Countless people deciding to burn the world a little bit at a time for a few extra quid. Even now the world's ruined, we're all still doing it.' She sighed. 'What do you even want money for anyway? What could there possibly be left to buy?'

Cooper shuffled uncomfortably. 'A guitar.'

Eryn blinked. Rage still coursed through her, but Cooper's words didn't add up in the angry math in her head.

'A guitar? What do you want a guitar for?'

Cooper shrugged. 'To sing songs.'

'What is there left to sing about?'

Cooper looked around wide-eyed. 'Snow.'

'Snow?'

'Yeah,' Cooper shuffled, 'and the way the ice gives the sun a halo. Or the hum the power lines make.' She patted the pocket which contained her notebook. 'I've been making up songs since I was a kid. I thought a guitar would help them sound better, but they're expensive. Doing this job, I could afford one. And maybe

I could practice during trips out to turn the turbines around.'
She hugged herself. 'Thought it might be a nice way to spend the time. Y'know, before the not surviving bit.'

Behind Cooper, darting cautiously across the snow was a small family of arctic foxes. A mother and a cub. Eyes as bright as Cooper's.

Eryn wiped her face.

Then she started to laugh.

Cooper jumped at the sound. The foxes disappeared into the drifts. Eryn's laugh echoed out over the cold, shaking her tense shoulders loose, creasing her face into expressions she hadn't made in years.

'What did I say?'

'Nothing. Everything,' said Eryn, rubbing the tears from under her goggles before they could freeze. 'Let's get back in the grub. We've got a long journey ahead of us.'

Cooper sat crossed-legged in the passenger seat, her notebook splayed out on her skinny legs. Her eyes were down on the page, but they weren't moving. They'd stayed there ever since they'd left the broken turbine.

Once they hit the existing ice tunnels and descended back into the frozen dark, Cooper cleared her throat.

'Do I want to know how I did?'

'Well,' said Eryn, 'we got one turned. So not bad.'

'We also lost one.' Cooper looked at Eryn warily.

'Yeah, we lost one. But we turned one.' Eryn leaned her head from side to side, as if weighing up the numbers. 'That's still one more than we had working for us at the beginning.'

Cooper brightened a little and turned back to the notebook in her lap.

Eryn swallowed spit into her dry throat. 'I'm sorry about earlier. When I lost my rag at the turbine.'

Cooper forced a smile. 'It's ok. When I told my mum that you were a bit, well, older, she said to be careful with what I said. She said you'd probably seen some shit... uh, I mean, stuff.'

Eryn gripped the wheel and sniffed. 'Your mum's not wrong. The world can feel very broken sometimes, and all I can see is everything I've lost. It can get the better of me. I shouldn't have taken that out on you.'

Cooper cocked her head. 'What if we could fix it?'

Eryn frowned but held her tongue. 'What are you thinking?'

'Could we use a lattice splint, a bigger one, to fix the broken shaft? We could use salvaged parts from the broken farm we saw.'

'You're talking about the turbine,' Eryn shook her head at herself. 'I thought for minute you were talking about the world.'

Cooper shrugged. 'Why not both? One turbine at a time, right?'

Eryn groaned at her own words being echoed back at her.

'One turbine at a time.'

As they drove back, Cooper hummed and Eryn allowed herself to dream again of loud music, of thrashing around a dance floor, of throwing up the horns one more time.

Together, they travelled in that small pool of light, steady through the cold and the dark, hoping to be home soon.

Rhiannon A Grist is a Welsh writer of Weird, Dark and Speculative fiction. Her work was included in NewCon Press' *Best of British Science Fiction* 2019 and 2020. Her novella *The Queen of the High Fields* (Luna Press, 2022) was longlisted for the BSFA Awards, shortlisted for the SCK Awards and won Best Novella at the British Fantasy Awards 2023
Twitter: @RhiannonAGrist Instagram: @rarrmageddongrist

The Leaves Echo What the Body Forgets

Marisca Pichette

My **grandmother's ghost** grows between the floorboards. Her spirit unfurls in soft white roots, creates a mat for me to walk on. She is cool, slow to spread.

I don't remember much about her. She was part of the last lost generation, rolled under rising seas and wildfires, overcome by grief for what had been. I know her death meant little in the grand scheme, the global shift that brought a species to its senses.

But walking her body, feeling her memory in my bare soles, I know no loss was meaningless.

I crouch, carefully trimming back dead roots, sculpting her shape into enduring health. My grandmother's ghost is long, flat, covering the ground level of the house. She requires more upkeep than the others.

I set her dead in my basket and straighten. Outside, the sun is struggling to rise. Clouds restrain its light, turn the fields bitter orange.

'May?'

A silhouette interrupts my view. Future leans through the open window, her scalp shining. 'Anything for me this morning?'

I set my basket on the table and cross the room to another ghost. Ivy climbs the wall, leaves red and gold. I know every tendril, but I run my fingers through, searching for fresh anomalies. From outside, Future watches, still as fossils long forgotten.

Art: Emily Simeoni

'I think…' My index finger catches on a withered leaf. I pluck it free. It's rolled closed, soft and leathery. I want to look, but this ghost doesn't belong to me. I pass the leaf to Future.

She exhales, peeling open a message from her past.

Behind her, the sun breaks through the clouds. Her outline is radiant light. My face warms.

'Anything?' I ask, trying to hold hope in check. Sometimes we are gifted innovation, stories, traditions nearly lost. Othertimes our ghosts pass us anger, sadness, regret. There is worth in both, but to survive, we need advice. And all those who carry it are gone, their memories encased in seeds and roots.

Future's eyes squint, her lip pinched under her teeth. She reads the leaf three or more times before finally smiling at me.

'It's for legumes. A new way of planting.'

A knot unclenches in my chest. Our legumes have been struggling to reach maturity in the sour soil. Gently, I correct her. 'You mean an old way.'

Her smile broadens. She presses the leaf to her heart, closing her eyes. 'Thank you, Grandfather Thrice-Lost. Your wisdom endures, and your descendants learn.'

My pulse stumbles as Future takes my hand. 'Thank you, May. You care for him well.'

Her gaze is bright, full of possibility. I'm warmed by more than the sun. 'We all care for those who chose to stay,' I say, conscious of the thickness of my voice. 'His kindness lives on in his line.'

Future laughs. 'I think you give me too much credit. I don't have anything for you, I'm afraid.' Her mirth dims, a flicker of uncertainty.

My aunts grow in her house. They knew much about husbandry, before time turned them into bark and heartwood.

'They've always taken their time,' I say, sliding my hand free of Future's. I remember their faces, their voices – reflections of each other, echoes of women lost to the past. All they once were lingers in me, irreparably altered. Are we less than our ancestors? Are we more?

I remember the ghosts growing in my house, waiting for me to tend them. 'Maybe tomorrow, or the next day.'

Future nods. 'In the meantime, I keep them green.'

I pick up my basket, feeling its roughness. 'And they keep us alive.'

After Future leaves to bring her grandfather's wisdom to the planters, I climb the stairs, mindful of my uncles clinging to the walls. I find two messages in their creeping vines, but no advice for developing our agriculture. Instead, warnings.

Any technology that powers by burning will eventually choke.

In a race against evolution, innovation will always fail.

I wind the creepers around my fingers. My uncles have always steered in this direction. Their information isn't new, but it is an evergreen reminder of what came before. While so many ghosts hope to steer our future, many focus instead on our past.

They don't stay to help us build. They stay to ensure we don't destroy.

At the top of the stairs, the light dims. My cousin grows outside, a willow draped in golden branches. Their leaves fill one window. I check each, plucking away dead ones and searching every vein for lessons. Today, they offer none.

Between their swaying branches I glimpse Future in the field, demonstrating her grandfather's technique to our best planters. My chest swells.

She seems to shimmer, soft at the edges. Her shadow feels more real than the rest of her. Watching, I realize the sun is directly overhead. I need to reach the rest of the house while I have the best light.

Bedroom, bathroom, attic. I tend the ghosts of ancestors, acquaintances, and strangers. In their roots and leaves, I find echoes of lives lived, ways lost.

Fish paste mixed in soil will fertilize.

Sing to seeds, and they will grow beautiful.

Break any weapon that grows stronger than its creator.

Never let currency become worth more than survival.

Live alongside spiders, for they will control pests that strive to devastate your harvest.

Some new information, much old. I fill my basket with wisdom and warnings as the sun crosses the sky. It takes me four hours to tend all the plants on the second floor. I eat fruit from a distant relative, drink water filtered by spirits unseen. As the afternoon darkens, I return to the ground floor, my feet finding my grandmother once more.

An anomaly. I lower myself to my knees, brushing my fingers across the message grown in her roots.

Don't make our mistakes.

Future returns after dark, knocking softly on the doorframe. I turn from where I'm stoking the fire, embers echoing spots in my vision.

'Well?'

Future comes to sit on the floor beside me, picking dirt from under her fingernails. 'We won't know for a few days. The planters are hopeful.'

I sit back, allowing that chance to relax my muscles. If the legumes grow well, the harvest could be double what it was last year.

'May?'

Future scoots over, resting her head on my shoulder. My skin prickles, heat sparking down my back. I open my mouth but can't taste any words.

She sighs in my silence, long and low. 'When they're done... when they've passed on all they know...do you think they'll leave us?'

'The ghosts?'

'The plants.'

I turn my head, my cheek resting on Future's smooth scalp. 'Not as long as we care for them. They'll keep growing.'

'And the messages?'

I stare into the fire. 'I don't know.'

Future shifts, fumbling in her clothes until she produces a brittle leaf. I recognize the kind – my aunts' species, twin trees that grow through Future's house.

'I found this, hidden in their roots.' She hands it to me. 'I'm sorry…I peeked.'

When I unfold the leaf, it breaks into fragments. Across dry veins, I find a message from the women who raised me.

Maybe—
Death is necessary
For in the arms of ghosts
New lives are born.

Future curls her arm around mine. 'What do you think you'll be?'

I let my skin touch hers, share the warmth. I think about our fields, struggling, but growing. I think about the clouds above us – once fever-red, now softer, more forgiving. Rain that tasted bitter quenches where it burned before.

We are recovering, with their help. Maybe my life won't see the final healing, but my descendants will.

'I'd like to be moss,' I say, so low I barely hear myself.

Future's grip tightens. 'I'd like to be the stone you grow on.'

'You can't be a stone. You have to grow.'

She shoves me, raises her head to look into my eyes. 'Then I'll be a tree, and let you cover me in green. I'll wear you like a blanket, and never be cold.'

She is more beautiful than any flower, any leaf. Any dawn, free of flames and draped in dew.

'I'd love that,' I whisper.

'Will you promise to write me messages?'

'Every day, and every night.'

Future lowers her head, weight returning to my shoulder. We

stare into the fire together.

'Then I'm not afraid of missing it,' she says. 'This – what's to come. Even if no one tends us, we'll have each other. Even if no one sings to our seeds, we'll trade secrets only we can hear.'

I feel the edges of my grandmother's roots, stopping just short of the fireplace. She died so long ago, yet here she is below us – growing, giving, hoping for a world better than she left.

'How long do you want to stay?' I ask.

Future shrugs against me. 'How long do *you* want to stay?'

I think about the fields, my ghosts, hers. 'Until everything is green,' I say. 'Until the sky is blue, and the stars bright.'

'Do you think we can do it? Do you think we can bring it back?'

I shift towards her. My aunts' leaf falls from my lap. The pieces cast long shadows in the firelight.

'With their help – and eventually, with ours – yes.' I hold Future in my arms, feel her heartbeat in my chest. 'All we have to do is listen. All we have to do is learn.'

Marisca Pichette collects plants and imbues them with ghosts. More of her work appears in Strange Horizons, Fireside Magazine, The Magazine of Fantasy & Science Fiction, Fantasy Magazine, and PodCastle, among others. She is the winner of the 2022 F(r)iction Spring Literary Contest and has been nominated for the Pushcart, Utopia, and Dwarf Stars Awards. Her speculative poetry collection, Rivers in Your Skin, Sirens in Your Hair, is out now from Android Press. They spend their time in the woods and fields of Western Massachusetts, sacred land that has been inhabited by the Pocumtuck and Abenaki peoples for millennia. Find them on Twitter as @MariscaPichette and Instagram as @marisca_write.

An Institution

Elis Montgomery

Elis Montgomery is a speculative fiction writer from Vancouver, Canada. She is a member of SFWA and Codex. When she's not writing, she's usually hanging upside down in an aerial arts class or a murky cave. Find her there or on Twitter @elismontgomery.

'Art studio, movie theater... Gardens! So many gardens. Maybe you'll get your green thumb going again? Restaurants, a little corner store...'

We've been through the specs a hundred times – Mom's list making isn't for our benefit. Grandma's charitable nodding is the rhythm to Mom's anxious tune.

As the descent pod lands in the centre of the village, Mom stands between us, holding our hands. I know she's thinking about the facility that hosted her own grandmother when her mind started slipping, all white walls and iron hallway gates. I squeeze her hand three times: our intergenerational signal of reassurance.

The pod's doors open, and we exit into a courtyard heady with the memory-promoting scents of sun-warmed herbs. Around us, sunlight sparkles on the burbling fountains, on the agromined metal frames of cottages with blooming ecowalls. A honeyed breeze carries birdsong and the soft chatter of strolling residents and plainclothes caretakers. In the distance, beyond the mosaic of gardens and cottages, I make out the wall of woven vines that cradles the village, letting the residents roam safely.

Grandma suddenly rips her hand from Mom's, rushing for the courtyard's cluster of carbon-sequestering chanterelles as big as oaks. We reach for her—

An old habit. She's safe here; she deserves what independence universal memorycare can give her.

So we let her go.

Grandma beams up at the chanterelle's folds as pollinators come to swirl about her head. Then she laughs – *actually laughs* – and starts naming butterflies. 'Monarch, morpho, swallowtail...'

I'm touched, but Mom shudders, sobbing silently beside me. When emotional list making won't do, I know it's bad.

'Mom. What's wrong?'

'Nothing.' She turns her smile toward me. The hand that was holding Grandma's is pressed against her heart. 'She squeezed three times.'

Beirut Robot Hyenadrome

Thoraiya Dyer

He's ours for one more day.

'Let's ride him to grandfather's grave,' Aisha says.

'Too risky,' Med argues. 'He breaks a leg, we won't get the money to *pay* for grandfather's grave.'

But even though Med's stronger than her now, this year for the first time ever – Aisha was shocked when she tried to push him into the horse trough like she'd done a thousand times and nothing happened – he doesn't jerk the saddle cloth out of her hands, or bar her entry to Good Sort's stall.

Mould spreads on the high arches. The mosaic floor hasn't enough straw to soak up all of Good Sort's mess. Aisha wrinkles her nose.

Then Good Sort puts his muzzle on the back of her neck and whuffs. She inhales his exhalation. Hot horse-life smell passes into her body. All his loamy, lathery, bloody, wildfire, cut-grass essence flows along with it, recharging her like a kite recharging a battery.

Aisha puts her face into Good Sort's shoulder and whispers, 'Grandfather.'

Towards the end, Grandfather Yehieh hadn't recognised Aisha. Called her Fatoomah, her mother's nickname, and praised God for healing her lazy eye. *Inshallah, your husband will come home from the fighting soon!*

Aisha's father hadn't gone to the fighting. He'd left the country with another woman and never looked back. Aisha's mother sold roses like bloodied velvet on grenade-pin-strewn street corners. She'd worn holes in her shoes almost as famous as the holes in Grandfather's fishing-line-laced boots. Valuable land doesn't buy shoes.

Asset-rich isn't rich.

At the very end, Grandfather called hoarsely, incessantly for his Aisha. But he'd meant his dead wife, not his granddaughter.

Aisha breathes out into Good Sort's nostrils.

Take my life in return.

Then, she knots a rope so soft her fingers can barely feel it over the horse's nose. Vaulting up onto his blanket-covered back, she grips the makeshift reins she's woven from the rope halter.

Med throws the stall wide open.

'You might be the oldest,' he declares, for the tenth time, 'but I'm the male descendant. The Sunni court will award Forest Stables to me!'

So I'll go to the secular court.

Aisha hates remembering their argument over Grandfather's dead body. Being on horseback makes her above all that, taller than any adult. She looks over Med's head, past the cracked, weed-seeded road.

Pine trees, their trunks jagged like lightning strikes, cage a garbage mountain in the middle of the hippodrome's well-compacted track. Robot hyenas scavenge single-use plastics, their recycled feet clacking on rocks or pinging on old benzene cans.

A hundred colourful kites, some on kilometres-long tethers to generate charge, others fluttering twenty or thirty metres above the pines, dip and strain overhead.

'Is it still a stables?' Aisha asks impatiently. 'Without horses it's just weird buildings. Yallah, coming?'

Good Sort never did get a proper racing name from Grandfather Yehieh. With no other horses to race against, why shouldn't their last horse go to the Emirates, to become Forest Stables Molten Mountain Snow King, and live in a country where the low human population density means they can afford to spend International Carbon Credits on horses?

'He's my property,' Med says. 'I'm not leaving him alone with you!'

'Whatever!'

Reeking of teenage boy-sweat and the pine scent he uses to try to cover it, he heaves himself over Good Sort's rump to sit behind Aisha. He doesn't try to take the reins.

'Let's go then,' he says stiffly, as if already steeling himself not to cry when they reach the cemetery, with its smouldering sacred herbs and shadows over gold-painted, engraved calligraphy.

If the robot hyenas on the garbage mountain were pigeons, they'd scatter, but they don't even look up as Good Sort canters past them.

'Grandfather won this land gambling on a horse race,' Aisha shouts against the wind of their passage. 'We can't agree on whose it is, now? Let's gamble on a race. The winner takes the stables.'

'Are you stupid?' Med shouts back. 'We're on the same horse!'

'Not a horse race. We'll race them.' She points at the hyenas. Old people sometimes take the plodding robots to carry their shopping home. Holding pieces of plastic in front of their sensors steers them. 'Three times around the circuit.' Grandfather never allowed the children to race. Said it was too dangerous. There was that time a taxi driver tried to run down the rider of a losing horse. 'The hippodrome can be a hyenadrome for just one day. Lebanese Independence Day.'

Good Sort tries to turn clockwise, to start the familiar circuit, to flatten into a gallop, but Aisha guides his head towards Peace Park.

'That's one week away,' Med yells. 'Why not now? Why not get down and race on our own legs right away? I'll beat you before you've even started.'

'Because,' Aisha snaps, 'any idiot can run in a circle. Modifying a robot hyena to race is a complicated job for smart people.'

Only a smart person, she thinks desperately, *can stop the tax men, the adults, loud and blind and selfish, from taking land away from two children.*

Anantha doesn't have time for a relationship.

Her mother warns her daily of the biological ticking clock to get her family started, but all Anantha feels is the clock ticking down towards the destruction of her country; its dissolution into hate-filled tribes sniping across a wasteland, buried toxins bubbling to the surface through repurposed civil war rubble corruptly veneered with marble and gold.

Minsa *does* have incredible eyes, though.

Incredible hazel eyes framed by unwaxed brows and a spangle of moles on nose bridge and temple that any self-respecting person should have frozen off in adolescence. Barely graduated, Minsa's work on kite generators saw her snapped up by Cedar Transformations, her prototypes set up as a demonstration at the golf club.

Minsa is a kite surfer, the Queen of Batroun. Those white beaches and aquamarine waters north of Beirut gave Minsa the inspiration to harness the winds.

For seven and a half hours the prototypes performed spectacularly, soaring four hundred metres in the sky, driving electricity generation as they unspooled. Subtly changing their sail shape so that they'd drift back to earth, they snapped their carbon fibre frames back into shape to soar once more.

But the golf course kites were too close to the airport, too close to Hizb headquarters, plus nobody thought to warn the Israelis about them. Had the drone that destroyed them been preventing surveillance, eliminating a technological unknown, or clearing a flight path for Ryan Bousaleh's private jet?

Minsa is undeterred.

'The old man lived under a bridge,' she says, the tiny coffee cup in her hand gone cold. 'He'd been the librarian when he was in jail, so even though he was sleeping in one cardboard box, shitting in another one, he'd started to collect piles and piles of second-hand books.'

'I heard about him,' Anantha says, scooping her majedra with a triangle of bread and the last pink turnip pickle. 'Someone set his books on fire, right?'

'I was there.' Minsa sets the cup down and spreads her hands flat on the table. 'Wahyetallah. Bayih's car broke down on the way home from Tyre. We tried to push it. Then we heard people screaming. I saw the orange colour on the bridge pylons and ran towards it. My parents started screaming. My brother ran after me.'

Anantha laughs. 'Do you always run towards fire?'

She wants to kiss Minsa's nose bridge. She wants to kiss her temple.

Minsa ignores the question.

'It was like something from a dream. Like the Great Library of Alexandria burning. People stood around staring into the fire, saying Allah yerhamou, as if it was a funeral and the old man had died, or as if the library was the body of a great king being cremated.'

'But the old man was alive.'

Minsa bends to her black satchel and pulls out something palm-sized and oblong. Anantha prepares to look at digital photos on a phone, but the object is a book, maroon-covered with yellowed pages.

'He was alive. A great king in a filthy puffer jacket. Leaning on a cane. Blue eyes streaming tears into a filthy, bushy beard. He saw me, alone, a little child, watching. He wasn't doing anything, ya Anantha. Not trying to save a single book. Just watching with the rest of us.'

'You saved one, ya Minsa,' Anantha guesses, even more enamoured to discover poetry in the heart of an engineer.

'I saved one. I saved this one.'

Anantha puts her hands seemingly-accidentally over Minsa's hands as she picks up the book, feeling herself charged by the feel of Minsa's beach-bronzed skin, as though Anantha is the rope being dragged by Minsa's kite.

The title is pressed into the cover, traced in silver ink: *Traditional Kite Designs of Mount Lebanon.*

Anantha flicks to the first chapter, where a line drawing labelled *tayara kasab* shows the hexagonal, bamboo-framed kite with trailing skirt of streamers. The opposing page, labelled *koubh*, shows the steps for folding the older kite design out of a single piece of paper, and tearing the square-spiral tail.

A tiny, pink rose is pressed between the pages, staining them.

'That figures,' Anantha says, closing the kite drawings safely on the rose again.

'That book fire was the most terrible, powerful thing I'd seen,' Minsa says, reclaiming the book. 'What's the most terrible, powerful thing you've seen?'

Anantha doesn't have to think very hard.

'I was on a ski trip,' she says, 'with friends from university.'

The café chair under her, the table, Minsa and the smell of coffee all vanish momentarily; Anantha feels surrounded by cold; by snow, by the bare sticks of terraced cherry trees drawing faded sermons on the mountainside.

There's the gravel road, and the silver Mercedes, trunk gaping, pulled over on the bend. There are bearded boys with shotguns, smoking, and the clink of chains, for that brief second, through the open passenger window as Sophie slows the silent electric Jag to negotiate the bend.

'The terrible thing that I saw in the mountains was a dubbae, chained to the back of a car.'

'That is terrible,' Minsa says. She shows no surprise that of all the human horrors, it's an animal one which has shocked Anantha.

That silken, white hyena coat, striped with grey calligraphy, just like the cherry boughs against the mountain. Those heaving, bloodied ribs.

And the boy, wearing plastic sandals in the snow. His ribs as thin and heaving as the dubbae's. Armed only with a mobile phone, between the wild animal and the smoking shotgun boys.

'There was a kid,' Anantha says. 'He was crying. 'I'll tell Baba', he kept saying. 'If you don't set it free, I'll tell Baba.' Then our car was out of earshot. I asked Sophie, 'did you see that?' but she didn't see. I don't know what happened. I doubt they set it free.'

'I wonder who his father was?' Minsa muses. 'Someone important? Or some corrupt big shot. Kullun yaani kullun.'

Anantha doesn't blurt: You *are important.*

It's ideas like Minsa's that will save them, not any useless politician. Ideas, and staying power despite the privilege to get out. The fearlessness not to run to a place where it's legal to love anyone. Where street libraries aren't likely to be lit on fire. Licensing their technologies in foreign countries will lend them the money to rebuild their own backyard.

'My design for the plastic recyclers,' Anantha says, 'until then had been dog-shaped. After seeing that boy, I was disturbed. I was haunted. I changed my robot to look like a dubbae. Both just harmless scavengers, mishhaek?'

Hyena robots to renew an open grave of a country.

Anantha blinks. The snow recedes and the café returns. She waves to order fresh coffee.

'You really ate all that majedra,' Minsa says, her little mouth screwed up in disgust.

'You don't like it? Lentils are more sustainable than meat.'

'It happened or it didn't happen, that an illiterate shepherd came to Beirut.' Minsa says mischievously, instead of answering directly. 'He inherited a surprise amount of money, and wanted to treat himself to a taste of luxury. Previously, he was so poor that all he could afford to eat was majedra. Upon reaching Beirut, he sat down at the finest restaurant, overlooking Pigeon Rock, but he couldn't read the menu. He was too embarrassed to admit this to the Garçon, so he simply pointed to the longest, most impressive-looking word that was printed there.'

'I know this one,' Anantha says, smiling despite herself. 'Majedra.'

Minsa giggles. 'The shepherd was disappointed, but he ate

everything on his plate. He noticed a wealthy couple, three tables over, sharing a platter of seafood. It had oysters, lobsters, octopus, raw kibbet samak with orange and lemon zest. Sea creatures he'd never seen before! When the couple finished eating, the husband cried, 'Garçon, encore!' – and the waiter brought another seafood platter.'

Minsa's giggles turn into tears of laughter. Anantha has heard the punch line, of course, but can't bear to interrupt. 'The shepherd,' Minsa gasps through her tears, 'cried, 'Garçon, encore!' And of course, the waiter brought him another mountain of majedra.'

Minsa's infectious laughter makes Anantha start laughing, too.

'It's not that funny,' she says.

'It is!' Minsa insists. 'I can't breathe. Let's get out of here.'

'And go where?'

'Somewhere,' Minsa says, catching Anantha's brown eyes with her hypnotising hazel ones, 'where I can put my tongue in your disgusting majedra mouth.'

Aisha's throat feels thick and her eyes sting.

Good Sort picks his trimmed feet up tidily as the Emirati groom leads him away from Forest Stables. The float's double-wide. Cool air streams through the gap in plastic flaps at the rear. The ramp is red-carpeted. Good Sort is royalty.

Four dog-sized drones, defending and surveilling the float, hover over the four corners of it. They make black, buzzing shapes against a cloudy sky threatening rain.

'Take care of him,' Aisha calls to the groom, though she'd promised herself she wouldn't say anything. The Emirati throws a grin over her shoulder.

Lifting on hydraulics, the red-carpeted ramp seals groom and horse inside the climate-controlled box, and lights come on everywhere around the electric vehicle. Aisha tries to swallow but the lump won't go down. Then the entourage of float and drones makes its way along the dirt track that once connected Forest Stables with the hippodrome, but now connects a bunch of weird buildings to a field of pine trees and garbage.

On their way out, the wind changes direction. The kites swing, fluttering, over the track, and the drones chop through strings and tethers alike with their bulletproof rotor blades.

Your aunty is a prostitute! Aisha swears in her head, making a small sound of dismay.

Ten or twelve paper kites, decorative camouflage with long tails, sail away on the wind. But two larger shapes, from higher in the sky, take more time and a longer arc to crash to the ground.

Aisha and Mohammed sigh heavily, together. Then look at each other. Aisha stifles a laugh.

'Here I was,' Med says, 'thinking that at least with the horse gone, we wouldn't have to pick up shit. Let's get the broken ones.'

Because the kites always break, when they hit the ground.

'Yeah,' Aisha says. 'My stables need power.'

'No,' Med says, '*my* stables need power.'

Their moment of shared feeling is broken. Her lungs can't get enough air and the tight muscles of her crumpled face hurt.

She forces her face to relax. Grandfather, *Allah yerhamou*, sent them to the best schools, before his dementia worsened. Aisha had aced robotics. Med showed talent on the athletics track. She definitely has the edge.

'I'll show you,' Med says, 'how good I am at complicated jobs for smart people. I'll have my new kite built before you even get there.'

Then they both start running towards the beach, past new solarglass-and-steel skyscrapers. Down the wind tunnel formed by the onshore Mediterranean breeze. Dodging honking EVs, oily mud puddles and rubble piles.

Aisha loses sight of her brother long before she reaches the series of steep switchbacks down the cliff face that takes her to the so-called beach. The red tile roofs of suburbs claimed by sea level rise form barnacled islands in a rubbish marsh. Robot hyenas grind shopping bags, syringes and straws in their patient jaws, sunk hock-deep in the tide, but they're not the recycling systems Aisha needs.

It's the floating recyclers – the nurdle-eating whales with

ferrofluid blood for microplastics collection – that excrete kite sail parts and rib parts that drift to shore. All Aisha has to do is run down to the closest heap of grey-looking sticks delivered by the tide.

Trash turned to useful components for anyone to find and use.

No matter how many bombs fall on the land, how many power lines are cut, how many drones stoop like rabid eagles to tear things down, the sea brings renewal from self-sustaining tech. Saltwater metal-mining sharks 3D-print both the simple wiring and heavy pistons for the robot hyenas, while the robot hyenas 3D-print the solar panels that power the sharks, using minerals and pollutants harvested from the water.

I don't know why I'm still hurrying, Aisha thinks glumly. *Med will be back at the Stables by now.*

Aisha ropes her bundle of ribs and sail together with a paced-out fifty metres of extruded kite cable that comes from nylon-net-eating shore crab recyclers.

Never mind. This isn't the real race.

Rain patters on the face she turns to the sky in gratitude.

It washes evaporated salt from her lashes to her lips.

By the time she gets home, soggy and sore – *it's not home, not without Grandfather and Good Sort* – Med's kite is assembled and already flying.

He pretends he doesn't notice her, panting and dragging her materials, because he's too busy doing carrot-on-a-string with the shiniest hyena. Dangling plastic garbage in front of its forward sensor, he leads it into one of the stalls, where a compressor waits to blast the grit out of its joints, and a rag sits soaking in machine oil.

'I'm naming her Homeless Aisha,' he calls out.

Aisha ignores him, setting her rib pieces out on the rain-wet grass to assemble. She doesn't need the newest, shiniest hyena in order to win what is rightfully hers.

She's got a much better idea.

Anantha, unspooling the thick cable of Minsa's new kite on

her flat apartment building roof, waves at the tiny figure down in the courtyard. It's Yehieh, the kid orphaned by the port blast twenty years ago.

He wears too-big boots, laced with fishing line – and wears holes in them running from one end of Lebanon to the other. A charismatic liar with an injured baby rabbit in his pocket and a chameleon in the crook of his elbow, he pretends to farmers that he has a truck for transport. Then he sprints with boxes of strawberries on his back through the night.

Since Anantha's smart-dubbaes started extruding oblong droppings of polybrick to use for building materials, Yehieh has started running crates of those all around the country. Recycled plastic is lighter than fruit, and it doesn't go off.

'I want a turn!' he shouts from the ground, his palms cupped around his mouth. 'I want to fly!'

'Come up,' Anantha calls back, but even as she does, she feels the sway in the concrete slab beneath her feet. The disturbance in her inner ear.

Howling dogs.

Rising dust.

Earthquake.

The Levant Fault's stayed quiet through most of the wars, and now, perhaps aggravated by neighbourly fracking for the newest offshore oil pipeline, it's awake, and angry.

The last thing Anantha sees clearly is Yehieh diving beneath the apricot tree. She knocks her glasses off in her hurry to wrap her right wrist in the cable. When she kicks the levers open that hold the spool clamped to the roof mounting, the kite pulls her arm out of its socket, and she screams. The ten-storey building collapses to rubble beneath her.

She's airborne. Flying.

It feels like dying.

'You can't go inside,' the casino bouncer, Abu Musa, tells Aisha. Then his brows lift as he recognises her.

'Ya Bint Fatima!' he cries, embracing her. He kisses her on the left cheek, then the right, then the left. 'Your grandfather, Abu Fatima, he's left you alone, eh? Allah yerhamou.'

'That's right,' Aisha says. She tries to hug him around the neck only, so he won't feel the tool belt strapped under Grandfather's leather jacket, black for mourning. 'We had to sell our last horse to bury Grandfather, Abu Musa. I know the museum's closed. I just wanted to touch his old trophies. Can I just stand next to the silks and the sashes, and smell the polish? You don't even need to turn on the lights.' *Or the CCTV.* 'I know my way around.'

'Ahlan wa sahlan,' he says, a tear on his cheek, swiping his security badge and cracking open the door labelled EMERGENCY EXIT ONLY at his back. 'Tfaddal! Take your time!'

Aisha creeps through the prefab hallway where she was once shunted with Med. Grandfather had to keep the children out of sight of the dignitaries when the renovated casino museum was reopened. Abu Musa had obligingly removed the batteries from the fire alarms. They'd played football in a haze of cigar smoke. The bounce of the football against concrete hadn't carried over the clinking of champagne glasses.

Most of the museum is dedicated to the history of the casino itself. Waxworks of Omar Sharif and Elizabeth Taylor leer in the red escape lighting.

But there, in a corner, is the framed photo of Grandfather in fishing-line-tied boots, mounted on a grey mare. A sewn banner bearing the arms of the Society for the Protection and Improvement of the Arabian Horse in Lebanon hangs over the wild-eyed, mildly alarming, taxidermied shape of famed sire Jewel of Byblos. The stallion had been known to stablehands by the less glamorous nickname Majedra.

Good Sort has some of Majedra's blood in him.

Behind the infopanels on horseracing, over a less alarming, more kindly-looking stuffed cheetah, hangs the relic Aisha seeks.

Beside that relic is another framed photo.

This one shows diaspora hydrogen fuel billionaire, Ryan Bousaleh, who sank a considerable sum into a cheetah endangered species recovery program, before it was discovered he was racing them in Beirut on a secret subterranean greyhound track. Betrayed by multiple appearances on his own extremely profitable livestreams, the rumour goes that he fled to the protection of the high seas. His cheetahs were rewilded and released.

Aisha absently pats the cheetah's head as she passes.

Any other country might have been too embarrassed by the situation to brag about Bousaleh's net worth in a museum, but not Lebanon. Its supervillains are as lauded as its heroes, provided they're wealthy enough.

And the training device that taught the cheetahs to run in circles around a track, the robot 'rabbit' which is a high tech version of the mechanical flag on a wire that lured greyhounds to race for centuries, is still there.

Perfectly preserved. Not a speck of dust.

Aisha unveils her wire cutters.

Fishing is banned in the Mediterranean, except on Independence Day.

Anantha sighs with pleasure.

The sun's out and the sky is crowded with kites. City slickers kick off their shoes to wade in the sea.

After hours spent working on the rebuild of Anantha's

earthquake-wrecked apartment building, fitting recycled tyres into place between extruded polyrok frames, Minsa somehow still has the energy to continuously cast and reel in her lure from the rock platform.

Anantha lies on a towel on her back, eyes thinly lidded beneath oversized designer shades, feeling sand and blown cigarette butts move across her skin and dreaming of her next project.

Like the hyenas, but submersible.

There's so much rubbish in the sea.

Anantha's shoulder never healed quite right. She can't swim any more. But maybe her robots can.

'Look!' Minsa cries, and Anantha sits up, squinting, already preparing herself to fake excitement at the capture of some luckless fish. She loves Minsa. You pretend to be excited for the people you love. Especially when they take you in after an earthquake makes you homeless.

It's not a fish, though. It's a young man on horseback, racing along the sand. As he rides, he trails a string of angry swearing like a koubh trailing torn paper.

'Yehieh,' Anantha calls, and he pulls up beside them, the grey horse's sides heaving.

'I bought her, I finally had enough,' Yehieh shouts, his wide, crescent smile like a stuffed sambousik. 'Her racing name is Earthquake From Heaven.'

'She looks beautiful,' Minsa says, dropping her rod, throwing her arms around the horse's neck.

And Yehieh's heart is obviously soaring, like one of the illustrations in *Traditional Kite Designs of Mount Lebanon*.

'She's a good sort,' Yehieh agrees.

The brain of Aisha's racing rabbit isn't exactly plug-and-play with the hyena body.

But close enough.

In the almost-light, as Lebanese Independence Day dawns, surrounded by the ghosts of a once-overflowing family feasting, Aisha lures one of the more agile-seeming hyenas into the house with her mother's old nylon bra, and uncovers the rabbit from under her pile of clean sheets that need folding.

One minor miscalculation is that she can't connect the battery up until it's time to race. This rabbit does one thing only. Aisha drags the heavy, inert hyena on a toboggan to the track. She's filed its plastic feet, to reduce mass at the extremities. No different to putting lighter, racing shoes on a horse.

Med's waiting, mounted on his gleaming hyena, holding a fishing rod with dangling garbage to steer by, shirtless in a jaunty tarbouche.

'Happy Independence Day,' he calls, turning his restless mechanical mount in circles.

'Yes,' Aisha grunts, her teeth gritted, hauling the sled into place on the starting line. 'It will be.'

'Did you break yours?' Med's eyes soften. 'Sis. I won't actually make you homeless. The paperwork's just a formality, isn't it?'

'If it's just a formality, why not put it in my name?'

When she mounts the hyena, there's no sharing hot breath. A group of shabbily-dressed men and women approach from the direction of the main road. Med waves to them.

'What are you doing?' Aisha demands. 'Who are they?'

'Witnesses,' Med says loftily.

'As if I'd trust your witnesses,' Aisha scoffs, even as she recognises their Jewish poet neighbour, Med's Shiite hairdresser, their retired Christian maths teacher, and two wizened, retired Dom jockeys that used to play backgammon with Grandfather.

Out of the poet's sleeve comes a stopwatch, held high, and the woman grins.

'One! Two!'

Aisha knots her waist and legs to the smooth carapace of the robot hyena with the soft rope that won't touch horsehide again. Slumped over the robot's shoulders, she grips one wire with a piece of basketball rubber, preparing to contact and then twist the two wires together. Her heart pulses in her throat.

'Three!'

Med covers his hyena's tailpipe, activating the sensor that alerts it when extruding polyrok is stuck and it needs to step forward faster. He brings the hooked garbage closer to its face, and lurches away while Aisha remains fiddling.

'Yallah!' the poet cries hopefully.

Aisha touches and twists the naked wire.

Abruptly, her chest flies violently backwards. The rabbit hyena runs.

The back of Aisha's head bangs against the base of its tailpipe.

Wind rushes in her ears. It fills the open hem of her tunic, wafting over her belly, chilling her breasts. She's exhilarated even as she fears for her life.

If not for the rope, she'd be dangling in the mud as they take the turn.

Aisha sees sky.

She sees Med's astonished, angry face from beneath, as her steed speeds past him.

Then sky.

Pine trees.

Earthquake-cracked stadium seats.

Sky.

Pine trees again.

The underneath of Med's stubbly chin.

Stadium. Sky. Trees. Stubbly chin. Stadium. Sky. Trees. Stubbly chin.

Tears wet her eyes. She closes them.

I've won the race. Forest Stables is mine. Now how do I stop?

Clawing at her body, the ropes, the hyena, she fumbles to locate the wires and untwist them.

Her monster stops mid-stride, so suddenly that it flips over its own front leg, twisting. The twist saves Aisha's life as she, and it, land on one side in the mud, her left leg protected by the gap between its heavy head and neck.

When she regains consciousness, Med has cut her free with his pocket knife. The crowd is helping her up. They cheer, laugh, and slap her back.

'What a ride!' they cry, but their voices fade, and all she sees is the fear in Med's eyes. It isn't fear of losing his inheritance, but fear for her. For her life.

'We'll put it in both our names,' Aisha croaks.

'No need for that.' Med holds her face in his engine-greasy, bony, teenage boy-hands. 'I gambled all the money I had on you.'

The witnesses start counting out wads of Lira, but they don't seem disgruntled.

'We can expand the gambling,' the teacher says. 'It'll be like old times.'

'My sons can ride,' one of the jockeys says.

'Who needs the cruelty of whipping animals,' the poet says, 'when we can have this?'

Med doesn't let go of Aisha's face to take the others' money.

'Are you hurt?'

'I'm not hurt,' Aisha says.

'Alhamdullilah. You sure raced a great robot. Forest Stables Homeless Med.'

For the first time since Grandfather died, Med cracks a crooked smile.

Thoraiya Dyer is an Aurealis- and Ditmar-Award winning Australian writer and veterinarian. Her (over 65) published short science fiction and fantasy stories have appeared in Clarkesworld, Analog, Fantasy Magazine, Apex, Podcastle, Cosmos, Nature, anthology "Bridging Infinity" and boutique collection "Asymmetry." Thoraiya's big fat fantasy novels in the Titan's Forest Trilogy are published by Tor books. A member of SFWA, represented by the Ethan Ellenberg agency, she is an avid hiker and arbalist inspired by wild spaces and the unknown universe. Find her online at thoraiyadyer.com or at https://mastodon.social/@ThoraiyaDyer

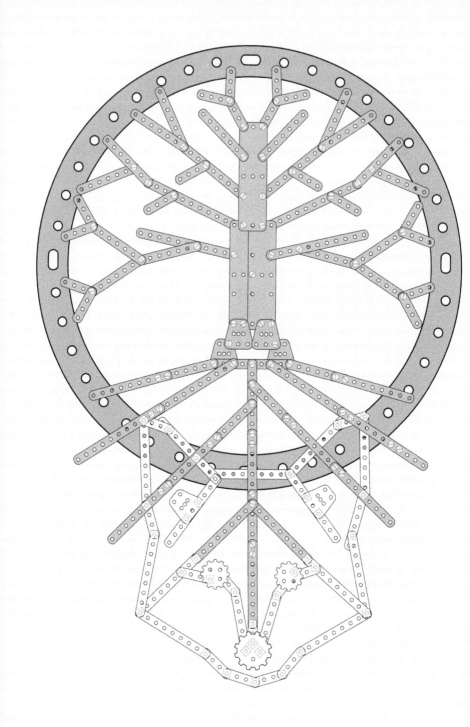

The Fox, the Hen, and the Green Hills

Gessica Sakamoto Martini

Whe**hen the first animal turns** into a human, Aya draws the hills behind her house in her sketchbook and colours them green – not dusty grey – because her grandmother once told her that, long before she was born, that was their colour.

When the first animal turns into a human, Aya looks at those hills and sees a white fox trotting on their barren soil under black clouds. Then, the fox stops and howls a song to an invisible moon. She sings until she grows taller, and her front paws turn into arms swinging at her sides. The white fox, now an old woman with silver hair, looks in the direction of Aya's house and smiles a smile of sadness and hope.

Aya tells her mother that an old fox woman is walking down the hills, but her mother dismisses her. 'One cannot be a woman and a fox at the same time,' she says, 'it must be the night robot that cleans the air'. Aya hates those robots and the drilling noise they make as they go around the town delivering food and water from the factories to every house and sanitising the air as they go. After they are done, the air smells like damp water, and it is impossible to stay outside. Not that Aya spends much time out of the house.

Aya lives with her mother and grandmother in a small cottage at the edge of town. Its walls are white like the rest of the buildings in town, 'to mimic ghosts and confuse death,' as her grandmother once said. But her mother had explained to Aya that it was simply the only barrier they could put up to reflect the sun's rays, save energy, and avoid heat. So, Aya only goes out before sunrise or after sunset, and sometimes, with her grandmother, visits the neighbours: an elderly couple who used to be botanists. They swear to Aya that behind the hills, where now only dryland remains, there was once a vast forest of oaks.

Nowadays, Aya's grandmother spends most of the nights praying. That makes Aya worry. Aya's mother asks if she is praying for a solution. 'I am praying for a remembering,' replies Aya's grandmother.

One night, Aya hears the white fox's song again, this time muffled, coming from her grandmother's bedroom. When she peeks through the crack in the door, she sees the old fox woman sitting next to her grandmother. With mouths closed, they sing a song that smells of soil. Her grandmother smiles a smile of sadness and hope. The following morning, Aya's grandmother is gone, and in her place is a hen laden with eggs roaming the house. Aya's mother cries for weeks. Aya eats the eggs and misses her grandmother for the first time in her life.

When the remembering spreads into town, people are afraid. Some recommend shutting all doors and not opening them even to the delivery robots. They talk of curses and the need to put earplugs in to avoid hearing the remembering. But Aya is sure that her grandmother heard it with her heart.

When the remembering spreads into town, not everyone is afraid. Some welcome the remembering, and soon it becomes difficult to distinguish people from animals. There is a dance between the two. Animals to humans, who now have a wild but gentle smile and walk as one with their surroundings. And then humans to animals, which Aya has never seen before: beavers, wild boars, honeybees, and small racoons, which roam the streets when evening falls. They are as grey as the hills, but somehow Aya likes them better.

Almost every day, strange tales are being told all over the town. A traveller from a coastal community tells Aya and her mother about people who have turned into whales and small fish that have turned into fisherfolk – with deep eyes and bodies that smell of seaweed – who, however, have put away the nets and ropes forever.

One afternoon, Aya glimpses a strand of silver hair coming out the back door of her neighbour's house. Now, her white cottage is shaded by two large oaks, and birds gather under their canopies. No one – thinks Aya – human or animal, has sung themselves into robots.

When Aya finally is able to go out for a few hours after sunrise, the hen disappears. 'Maybe a fox has caught it,' says her mother. 'Maybe,' says Aya, 'or maybe they now sit together as equals, breathing fresh air.' Aya's mother stays with her until Aya is old enough to buy her first canvas. Then, her mother sings herself into a hare and leaps into the night.

Aya now paints hills with foxes, hares, and hens dancing and singing under the full moon. She hangs the canvases all over the town. The buildings are now the colour of a rainbow.

One winter day, the old fox woman sits beside Aya on her bed. It is snowing outside. Aya asks her if she is a human, an animal, or both. 'There is no longer a difference, there never was, and there never will be,' the old fox woman replies. Then, Aya and the old fox woman sing together an old song of hope. Their smiles know no sadness.

The following spring, in the green hills behind the white cottage, a red fox is spotted running under the warm sun while a robot sits leaning against an old white wall, dust covering its body.

Gessica Sakamoto Martini holds a PhD in Anthropology from Durham University (UK). Her work has appeared or is forthcoming in FlashFlood (National Flash Fiction Day), Corvid Queen, Seize the Press Magazine, Crow & Cross Keys, and others. She is a Fiction first reader at Orion's Belt magazine, and can be found on Twitter at @GJMartini talking about fairytales. She currently lives in Italy.

Shelter from the Storm

Andrew Knighton

Just looking at the spiders made Mav tense, throat tight and stomach crawling. It didn't help that the spiders themselves seemed unwell, their bodies drooping onto the mycelial mass as it disintegrated from around their webs.

Through a gruelling effort of will, Mav made herself catch the spiders, holding the scoop out stiffly and precisely, tipping one after another into their slender cylindrical carriers. Every tiny twitching movement, every furred leg protruding from the collapsing fungus caught her eye. This was why spiders were her job: her intense awareness of their every movement, her inability to look away when one might be nearby. It was exhausting, and it was vital.

The fungus walls of the houses had crumbled in a desiccated drift into the dirt. All around, the clan were packing their belongings into trailers, hitching solar batteries to engines, checking the chains and tires of bikes.

'Are you ready?' Dev looked anxiously from the clan's precious tablet to the sky to the tablet again, carefully tapped a finger against the ancient screen scarred by generations of use, and slid the device into a pouch. Mav took a deep breath as she caught her tension spilling from the spiders over to him, and bit back her first response. She looked after the spiders, Talia looked after the livestock, and Dev looked after the people.

'Almost.' She scooped the last spider into its travelling home, snapped the case shut, and fixed it onto the back of her bike. There was a hissing inside the case, as its mechanisms lulled the spiders into something like a coma. One day, she would work out how the ancestors had made this thing, and she would build another one. For now, she was just happy if she could repair it when something broke. 'Now I'm done.'

She shivered. The storm clouds Dev had been watching roiled in from the north, threatening to pound bodies and soak paths, perhaps even to melt rocks if the winds came in the wrong way. The unexpected storm two days back, and forbidding frost that followed, were warnings the clan couldn't ignore. The sheep might want longer feeding on the rich glass of the Hoy, but they had to move out.

Dev took the lead, and Mav flung all her energy into keeping up with him, cycling around muddy puddles and across expanses of smooth old rock, finding a path for the others to follow. Racing Dev felt good, even if he didn't know they were racing. Watching for rough ground kept her from fretting about the spiders, too. Behind them, the clan rattled, clanked, roared, and clopped across the hills, heading south.

The place where they hid from the midday sun had once been coastline. While the clan sheltered under shades of the lightest cloth they could weave, and their batteries basked in the heat, Mav cracked open her spider case. It was like picking at a scab, uncomfortable but irresistible. She flinched as the clasp clicked open.

One of her passengers was moving in the tube, sluggishly spinning erratic trails. Did that mean the container was damaged, or that the spider was? Which would be worse? Mav frowned at her own meaningless comparison. How could there be a better choice when either spelled doom for the clan? Without the spiders spinning their webs, there would be no framework for the mycelium to grow up. Without that, there would be no travelling homes. They'd be forced to settle in one place, whether caves, crumbling cities, or something of their own construction.

No escape when ice froze the land, or the sun boiled the streams. A few of the clan might survive the first year, but they relied on each other, and every single loss brought their fragile community closer to collapse.

She clicked her fingers, breaking herself out of the spiralling thoughts.

'You OK, Mav?' Dev crouched beside her, speaking in his soft, steady voice. She hated when people tried to reassure her. She wasn't so fragile that she needed to be safely contained, but if she went silent or snapped back then it only got worse.

She could tell him about the spiders. If it was a real problem, then she should. But then he'd have to manage the situation, either soothing her or telling the others, and until she was sure, she didn't want to face either response.

'I'm fine.' She pointed at the small, coloured shapes protruding from the sand, fragments of a broken rainbow. 'Look, fish crap.'

Dev laughed. It was a name they'd given the shapes when they were young, because they appeared most often along the coast, or in the vast middens full of strange waste that the ancestors had left behind. Erina insisted that they were figures of people, portable icons the ancestors carried from one camping ground to the next, but Mav wasn't convinced. Sure, some had what could once have been arms and legs, before time abraded them, but others were squares, strips, strange shapes that couldn't have mirrored life.

'We'll have to take a detour,' Dev said, and Mav's shoulders unknotted as she realised that the soft voice wasn't for her sake, but to keep from alarming the others. 'One of the clans left a message, the Counting Bridge has finally rusted through.'

Mav's eyes went wide. 'Without the bridge…'

'The route south's twice, three times as long. We've got to ride hard to stay ahead of that.' He nodded north, to the bulging clouds. 'Good chance we get caught under the front edge tonight, but we can't slow down for proper shelters. If we stop at dusk, can you grow some emergency domes?'

Mav thought of the sluggish spider weaving its broken web. Maybe it was just that one, disturbed by her opening the case.

'I'll do it,' she said, with a false reassurance Dev usually saw through, but which slid past his attention, torn as it was between the clouds and the clan.

'Great. The moment it gets cool enough to ride, we roll out.'

There was a sense of unease in the clan as they made camp that night. This wasn't their usual route, and Dev had been drawing them on fast, but no one wanted to ask questions in case they didn't like the answers.

At least the spiders seemed better. Mav tipped them out, a few at a time, into the rings of mycelial cells and plant feed that others were laying out. Within moments, fungal flesh started to rise and the spiders laid threads from one grey cap to the next. Trained clan members coaxed the spiders upward and inward, their threads and the mycelium producing self-supporting frames that would become low domes within an hour and thickene into shelters within two. Not much space and not terribly sturdy, with hours of growth instead of days, but enough for a night.

Mav crouched by a river that ran through the rolling green, and cleaned out the spider tubes. The ground here was strange, turf breaking underfoot in odd places to reveal pools of pale sand underneath. In the river's shallows lay clusters of dimpled spheres, their white shells stained with age. Mav picked one up and rolled it between her fingers.

'Never seen a golf egg before?' Dev asked, squatting beside her. Did he have to creep up like that?

'What's a golf?' she asked, her voice a saw edge.

'According to Erina, they're a species of ancestors and these were their eggs. The shells are extra hard so that they could hit them with sticks, knocking their young ahead of them as they migrated.'

'Erina's crazy.' Mav tapped a golf egg against the spider carrier. 'No way this could hatch.'

'It's a nice thought though, isn't it? That not all of the ancestors lived settled in their cities, that they were like us.'

Dev had that faraway look that made her want to reach for his hand, but to do so would bring back the soft voice.

'Mav!' Talia came running. Her foot sank through one of the

soft spots and she tripped, went sprawling, scrambled back to her feet. Dev ran to meet her, and Mav followed, clutching the transport box tight.

Not the spiders, not the spiders, not the spiders, she chanted over and over in her head, even as she stiffened in anticipation of their twitching legs.

'They're not weaving right,' Talia gasped, wide eyes aimed straight at Mav. 'What do we do?'

'Show me.'

They hurried to the nearest shelter. Instead of a low dome, the mottled grey flesh was growing at erratic angles, turning in on itself in some places, in others folding stiffly out like a broken egg. The next one over was forming a dome, but full of holes. Around them, clan folk spoke in hushed whispers, some looking expectantly at Mav, but more at Dev. She couldn't tell which was worse, the weight of expectation or the judgment of disregard.

While Dev reassured people, Mav examined the spiders. Some were sluggish still, laying thick layers of web over the same places over and again, encouraging the fungus to cluster in thick, useless lumps. Others ran twitchily from place to place, their webs full of gaping gaps.

'What's happening?' Talia asked.

'I don't know.'

There was something especially spidery about those twitching movements, bringing the dreadful possibility that one of the spiders might leap and land on Mav's skin. She stood back, her own eyes darting, trying to watch them all while she held herself back from their reach.

'What is this?' Dev asked.

'I said I don't know!'

For a moment, Dev's gentle smile vanished. It was the most dispiriting thing Mav had ever seen, like watching ancient land swept away in a deluge. When the smile returned, it looked to her as fragile as any web.

'It's OK, Mav,' he said, and laid a hand on her shoulder. 'No one's blaming you, but we would like your insight. You understand the spiders better than anyone.'

That voice, so horribly like an embrace, a voice that said she belonged. But Mav had family once, before a flood took them, and she couldn't take another loss like that. The step she took away from him brought her closer to a failed shelter, and she turned her red face to examine its disharmony.

This was her fault. She was responsible for the spiders. If she didn't keep them well, then who would?

'The spiders are sick,' she said.

'Sick how?'

She didn't know, and if she said that then he'd try to encourage her, to coax out an answer. She leaned in, watched one of the spiders wobbling between rising mycelial strands. Something about the way it moved, she needed a closer look, but she'd left the case with its scoop back where Dev was standing, and she didn't know what he would do if she came close. Would there be anger, disappointment, or worse yet, an understanding she didn't deserve? The spiders scared her less than him.

It took all her strength to reach out a single finger, lay it on the spongey fungal flesh, and let a spider climb on. Its legs were needles against her skin. Her finger trembled as she dragged it in front of her eyes, fighting the urge to scream and shake the creature off. The spider lay still, slumped, its belly pressed against the ridges of her skin, one grotesque limb twitching. It reminded her of a kid sick from eating bad berries, or of Erina after she'd been drinking whisky and emerged through singing and shouting into the maudlin, nauseous fever that followed.

'They've been poisoned,' Mav said. She looked past the faltering shelters to the approaching clouds and felt a sickness worse than spiders. 'The storm. It must have passed through the edge of a chemical zone, caught something that soaked through the fungus.'

'Can they still build?' Dev asked.

'Not now.'

'But soon?'

'I don't know.'

She felt relieved when he didn't ask her to guess, but then she wondered, how bad did things have to get for Dev to stop encouraging her?

'They won't build tonight,' she said.

'Will they live?' This wasn't the soft quiet of comfort anymore. It was the sombre quiet of stones.

She didn't know the answer. All she could do was keep the spiders safe and give them time, like a fever patient. Right now, the clan were the ones who needed caring for, and something in Dev had faltered, like a bike chain fallen loose off its gears.

'They'll live,' she said firmly. 'But if *we* want to live then we have to move, before those toxic clouds reach us.' She clapped her hands and the whole clan looked her way. 'Cut the fungus at the base and give it the dying dust. I'll be round to collect the spiders, then we ride out.'

'It's almost dark,' someone said.

'Hook lights on the vehicles. Put those without lights in the centre of the column.'

'The batteries won't last.'

'Then we ride until they stop.' She made her voice hard and loud. 'We have to get ahead of that storm. Everyone move. Now.'

She knew that she wasn't leading right, that they needed to be coaxed and encouraged, not shoved around. But she wasn't a leader, she was a desperate woman filling a gap in their web.

'Go on,' she said, lowering her voice for Dev. 'Get ready to ride.'

For a moment, he didn't respond. Then he blinked, nodded, began to move.

'Right you are, Mav.'

The bright light they'd tapped into at noon had given them enough power to travel through the night. It was nearly dawn before too many lights flickered out and the clan was forced to a stop. They'd reached the outskirts of an ancestor settlement, shelters made from square red rocks. Most had fallen in, and few had roofs, but by stretching sun shelters between the walls, the clan constructed somewhere they could wait out a few hours, as long as they didn't get a storm wind or heavy rain.

After the ride down from the river, Dev seemed more like his usual self, offering suggestions and assistance to those building the shelters, easing the process of raising their protection while Mav checked on the spiders. Usually, she dreaded opening the case because she might see the spiders move. Today, it was because she might not.

The spiders were mostly motionless, as they should be after a long ride. Near the top of the stack, a few twitched in their tubes. One reached up to touch the glass, as if to reassure her that he still lived. Mav swallowed, stretched out a finger, and returned the gesture, the glass guarding her from feeling that prickly flesh.

'You'll live,' she said, whether to herself, or the spider, or their torn and toxic world, she didn't know.

Footsteps approach across moss-strewn rubble. Dev, Talia, Erina, the leaders of the clan.

'Do you have a few minutes, Mav?' he asked.

'Minutes? Sure.' Exhausted from the ride and the effort of leading, she couldn't hold back bitter laughter as she waved at the sky, its pre-dawn grey hiding the catastrophic clouds. 'A year, though? Weeks? Even days? I wouldn't count on that.'

She snapped the lid of the spider container shut, then realised that she still held a tube. Rather than open the case again, revealing hundreds of tiny bodies, she slipped that one container into her pocket.

Dev led them away from the broken buildings, down a road whose smooth black surface had been shattered by some disaster, then smoothed by years of wear. A web of moss and wilted dandelions led them to the foot of a hill where they stopped and stared.

This had clearly been one of the ancestors' hidden middens, its indestructible waste buried in shame. The dirt covering it had been washed out, exposing tattered edges of shiny black cloth, jagged spikes of rusting metal, mouldy foam, and things Mav couldn't even name. The wind and rain that had exposed the remains had washed away the lighter parts, leaving a loose ring of small objects around the base of the hill, their bright colours emerging from the gloom of mud as the sun rose.

'Fish crap,' Mav said, staring at them.

'Travel icons,' Erina said.

'Plastic toys.' Dev held up his tablet, one of their last working artefacts from an earlier age. 'There are documents in here, some from the ancestors, some notes left by the clan's leaders. I read about when they made these toys, billions of them, more than all the children in the world.'

'Why didn't you tell me?' Mav picked one up, a black piece that could once have been a spider, when looked at in the right light. Its legs were limp and twisted, two of them reduced to stubs.

'Would it have made anything better?'

Mav turned the toy in her hand. Did it seem ordinary because Dev had wiped away the mystery, the endless possibilities, or was she too tired to care?

She hated it when he was right, when he knew things she didn't.

'Will the spiders recover?' he asked, grim-faced.

Now was the time for truth. 'I don't know.'

'So, what do we do?' He turned to the others.

'We could make other shelters,' Talia said, her voice holding more hope than certainty.

'From what?' Dev asked. 'Do we have anything else that's as sturdy, and that will go up quick enough?' No one responded. 'Well, it's one option. What else?'

'Settle somewhere,' Erina said. 'Caves, maybe a city. Turn it into a home where we can huddle down and live through the weather.'

'Wherever we go, we'll be too hot sometimes, or too cold.' Talia shook her head. 'Both, maybe. Not everyone can live through that.'

'Some of us will.' Erina had lived through times when the world was even more wild, when the clan way of life couldn't save everyone. The world was healing, and if it cared better for people, it did so out of indifference, not love. 'Better that than risk losing everyone.'

'There has to be something,' Dev said, trying to be his usual self, to reassure and encourage. This weak version was worse than nothing at all.

Unable to bear the sad weight of their conversation, Mav turned away. She picked up more of the plastic toys. They were so light, they felt like they should be fragile, but when she pressed a thumbnail across one it barely left a mark. How strange, that these would survive the end of the world, but people wouldn't. She hooked one through another, then another, making a chain, as jumbled and distorted as the spiders' sickly webs. At least she wasn't poisoned like they were, so she could fix her own work.

She laughed, the first joyful sound of a dark night.

'Fish crap,' she said. 'Grab as much as you can.'

She scooped up an armful and rushed back along the road. On the horizon, feverish clouds billowed across the dawn.

At Mav's excited shout, people emerged from their shelters. She shouted instructions, and one of them laid out a ring of mycelial spores, retrieved from earlier failed attempts. Another poured on feed, and a third water. The fungus started to spread.

Without spiders crawling across it, lending structure and support, the fungus seeped out across the ground. Mav took a moment to muster her hopes, then thrust a plastic toy in, the imitation spider she'd first picked up. The fungus rose around it. As it peaked, she added another toy, and the fungus took hold of that, then the next, and the next, forming an arch around the pieces. Others followed Mav's lead, placing plastic in the fungus, shaping it as it rose. A shelter formed, lumpy, irregular, its surface broken by brightly coloured protrusions, but a shelter still.

Dev, Erina, and Talia handed out more plastic. Someone laid out another ring, and another. Emergency shelters, small domes to survive a storm, sprouted across the broken ground.

'It's not as good as the spiders,' Mav said, 'and it'll be harder to shape right, but the plastic's light, so we can take it with us. Plus it takes up space in the mycelium, so the shelters will grow quicker.'

'Will it survive a storm?' Talia tapped the first shelter. It wasn't as sturdy as Mav had hoped, but it held.

'Enough will,' she said, and for once she was certain.

She found the spider case and set it down in a sheltered spot, surrounded by a small ring of mycelium. As the fungus rose, she fed it plastic, building thick, sheltering walls to keep her wards safe. Perhaps some wouldn't survive the storm's poison, but she had been watching them for years, every terrible, beautiful, twitching move, and some would survive, with her help. If the clan wanted the spiders to look after them, they had to look after the spiders.

Dev knelt beside her, carefully feeding plastic into the fungus.

'Thank you,' he said quietly. 'The clan needed me last night, and I…'

She laid a hand on his arm and he turned to smile at her. It wasn't the smile he gave others, soft and reassuring. It was a bolder smile, one only for her.

'We do what we have to,' Mav said. 'You, me, the fungus.' She tapped the grey flesh hardening in front of her. Then she took a tube out of her pocket and peered at its occupant. Was it her imagination, or was he looking better, raising two legs now in greeting? 'Even the spiders.'

'Are you going to free him?' Dev asked. 'A reward for his help?'

'And risk him crawling over me in the night?' Mav laughed. 'No chance.' She smiled. 'But I might not hold him at arms' length anymore.'

Andrew Knighton is an author of short stories, comics, and the fantasy novellas Ashes of the Ancestors and Silver and Gold. Working as a freelance writer, he's ghostwritten over forty novels and hundreds of pages of content marketing. He lives in Yorkshire with an academic, a cat, and a heap of unread books. You can find him at andrewknighton.com and on Mastodon as @gibbondemon@wandering.shop.

The Best Taco in San Antonio

Christopher R Muscato

Leyes de Vegancia: 6:45am
The city awakens slowly; the pace of life in San Antonio is never idle, but rarely hurried, and less so in the morning. But as the chimes of metallic roosters shake the city's denizens from their beds, Ignacio is already scrambling eggs, sautéing peppers, and warming tortillas de maíz. In photovoltaic heating pads and parabolic saucers, he captures the power the Sun, transforming it into the other forms of energy that sustain this city: coffee and breakfast tacos.

The plaza is soon abuzz, and Ignacio has a line stretching from his taco truck nearly as far as the fountains by the time the trolley bells signal the start of the morning commute. Some risk running late to grab a bite from *Leyes de Vegancia*, but what's a missed conference call compared to potatoes roasted in wild sage, the fluffiest eggs this side of the Rio Grande, and a generous distribution of peppers, wrapped in tortillas still steaming with heat from the griddle? And that doesn't even take into account the salsa, because of course there is salsa. Several kinds, in fact.

As the morning rush wanes and the sun rises higher, Ignacio packs up his photovoltaic pads and saucers. The multi-leveled truck meanders towards its next location, chasing sunlight through the city. Pausing at an intersection, Ignacio sends a quick message.

Art: Andrew Owens

Battle of Flowers: 9:45am

Ruby turns from her station and gives Margarita a peck on the cheek. Margarita blushes and swats her partner away, then returns to chopping onions as Ruby dips out of the kitchen to respond to a ding from the comms. Of San Antonio's famed solar-cooking food trucks, their Louisiana/Filipino fusion is one of the most unique.

'What's up?' Margarita calls out.

'Old man Ignacio.' Ruby reappears in the kitchen. 'He's looking for five-point-star jalapeños.'

Margarita sets down her knife.

'Did he say what for?'

Ruby shakes her head and bites her lip. Margarita runs a hand through her hair.

'Okay, let's see what we can do.'

Alamode Tex-Mex 11:05am

'Here ya go, darlin',' Rosemary scoots a bowl of chili through the window. 'You want cornbread with that? What am I sayin', course you do. You wait right there.'

Without giving the customer a chance to respond, Rosemary ducks below the counter.

'Fresh from the solar oven. Hotter'n a carbon sequestration officer's temper. Careful with that, hear? Alright, outta my line, I've got other customers.'

As the frazzled customer scuttles off, Rosemary peers around the interior of her mission-shaped double-decker food truck.

'CESAR! Where…?' She pulls herself just far enough up the ladder to see her husband rummaging through the greenhouse. 'The hell you doing up here? You see this line outside?'

'We got any of those five-points growing in here?' Cesar yells through the greenery.

'The pepper? Seeds from the last batch never sprouted. Why?'

'Ignacio needs one.'

There is a moment of silence, then Rosemary's voice rings from the kitchen below.

'Y'all heard me, truck's closed! I don't care how long you've been waiting, just find us later with that food truck app! Alright now, scoot!'

Leyes de Vegancia 12:02pm

Ignacio glances at the clock. The vegetarian fideo loco is stewing nicely in the solar cooker, but the rest of his kitchen is filled with half-chopped ingredients, recipes unfinished. A few regulars of *Leyes de Vegancia* notice the chaotic state of things, but hold their opinions.

There weren't any five-points in the community gardens that morning, although Ignacio did encounter a tangled lattice blown over by the winds last night. Fixing it took time he didn't have, but it needed to be done. He isn't the only one who relies on those gardens. At least he was able to collect a few eggs from the free-range chickens.

Ignacio checks his solar-intensity map. There are a series of pocket gardens along the riverwalk where peppers might grow, and the sun will soon be in a good position for him to start roasting vegetables there. A rare delicacy, devilishly hard to cultivate indoors, the five-point-star jalapeños almost always had to be foraged.

Ignacio is packing up when there's a knock on the window.

'I'm sorry sir,' a young man tips his hat, 'But I'm new to the city and I missed my trolley. Can you direct me to a D-line station? I'm trying to get back to my office downtown.'

Ignacio rubs his chin.

'There's a skytram across the park but it's a long walk if you're in a hurry. Hop in, chico. I'll drop you off.'

The young man removes his hat as he enters the truck, overflowing with thanks. He says he hopes it's not too far out of the way. It is, but Ignacio doesn't say so.

Alamode Tex-Mex 2:15pm

The food truck *Alamode Tex-Mex* jumps along the road like an unbroken bronco, pots clanging in a discordant symphony.

'I'm telling you, it wasn't Hemisfair orchards. We found them at the Highline wind farm!' Rosemary shakes the map. Cesar shakes his head.

'You old loca, you don't know what you're talking about.'

'She'll be here soon!' Rosemary snaps. 'Just head to the Highline.'

Cesar jerks the wheel, the tires squealing.

'What'd you do that for?!'

'You said to go to the Highline.'

'You don't take Mission to get there, you take Federal!'

'Ay, you old loca!'

Battle of Flowers 5:28pm

Margarita is silent, eyes on the road. Ruby scowls. Yes, it's true that Margarita handles the finances since they merged their businesses, but Ruby knows they lost revenue today too. She knows how to run a business. Her soul-food truck did just fine when she was on her own back in Louisiana, thank you very much.

Margarita's stress is dense as August humidity, suffocating. With every mile, it grows thicker, churning with the weight of the business they lost today in pursuit of a single crop. Ruby doesn't know how much longer she can stand it. She tries to release the frustration twisting in her shoulders by taking a slow, deep breath.

That's when a smell wafts in from the kitchen. Ruby's grimace softens.

Her gumbo was always good. But it never had that unexpected, wonderful zing before adding a Filipino twist. And Margarita's lumpia was better now that it had a bit of Ruby's Cajun flair.

Moving in together, merging businesses, wasn't without its challenges. But look at what they had made. And it would have been much harder had San Antonio's most respected solar-chef not welcomed them into the community, giving them assistance, support, even cooking tips along the way.

'We owe Ignacio a lot,' Ruby places a hand on Margarita's shoulder. It's a simple sentence, and a simple touch, but the tension begins to instantly melt away, like butter in a cast-iron skillet under midday sun.

'We do,' Margarita leans into Ruby's hand. 'Hopefully we can repay some of that today.'

Ruby kisses her on the head as their truck rushes onward.

Leyes de Vegancia 7:35pm

Ignacio examines the pepper. It's small, not quite ripe, probably won't have quite the right kick, that distinctive flavor she loves. But at least he found one.

Knock. Knock.

'Abuelo!' A small girl flings herself inside as the door opens.

'Niña! You're early!' Ignacio exclaims. He wraps the girl in a hug, then looks through the door at the woman standing outside his truck.

'Hi Dad.' The woman says.

'Mija,' Ignacio breathes. He feels the tears already forming in his eyes. 'Welcome home, mi vida, welcome home.'

'Abuelo, is it true you live in your food truck?' The little girl bounces in front of her grandfather. He laughs.

'Yes, of course, all solar-cooks do. Come, I'll show you.'

And with that he whisks her on a grand tour of his double-decker truck, the convertible greenhouse/studio on top, the kitchens below where, years ago, he taught his daughter everything he knows about cooking, minimalist living, gardening.

'In honor of your visit I am making your mother's favorite taco,' he tells the girl. Heat rises along his neck. 'But I'm sorry, it took me too long to find the five-point-star jalapeño, and I could not find one that was very good.'

Before another word is said, there is a loud honking and a rattling food truck shaped like a mission church screeches into view.

'ALBA!' A woman screams, leaping from the truck and wrapping Ignacio's daughter in a bear hug. 'It's been too long.

I figured you were coming home when your father asked for peppers. Here, Ignacio, we found two five-points for you, don't ask me where, just get cooking.'

Ignacio fumbles with the peppers and Rosemary turns her attention to the little girl.

'Look at you, precious. You know, your momma practically lived in my truck when she was your age. Then when she was older, she babysat my little boy.'

'How is Clem?' Alba asks.

'Off at college, can you believe it? Says he wants to open a food truck in Denver, of all places.'

The conversation is interrupted by the arrival of another truck, painted with flowers.

'Sorry we couldn't get these to you earlier.' With a hug, Ruby hands Ignacio a bushel of peppers, a full day's work of urban foraging.

As Margarita introduces herself to Alba, a blushing Ignacio takes stock of the bounty.

'This is too much for three people,' he declares. 'Please, friends, stay and celebrate with us.'

Ruby helps Ignacio prep ingredients while Margarita and Cesar set up a grill with clean-burning biofuel logs, and Alba and Rosemary set up tables.

As the sun sets on San Antonio, generators hum to life, redistributing the day's harvest of sunlight throughout the city. But on this evening the keepers of the food trucks are not among those settling in. Together they eat and drink, sing and dance, howl at the moon and dance under the stars as the day turns to night, and the world prepares for another day.

Christopher R. Muscato is a writer from Colorado, USA, the former writer-in-residence of the High Plains Library District, and a Terra.Do certified climate activism fellow. His climate-focused fiction can be found in House of Zolo, Solarpunk Magazine, and several solarpunk/hopepunk anthologies.

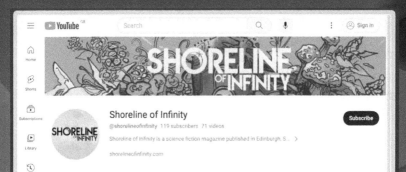

Shoreline of Infinity's Event Horizon: live, online and on Youtube.

www.youtube.com/@shorelineofinfinity

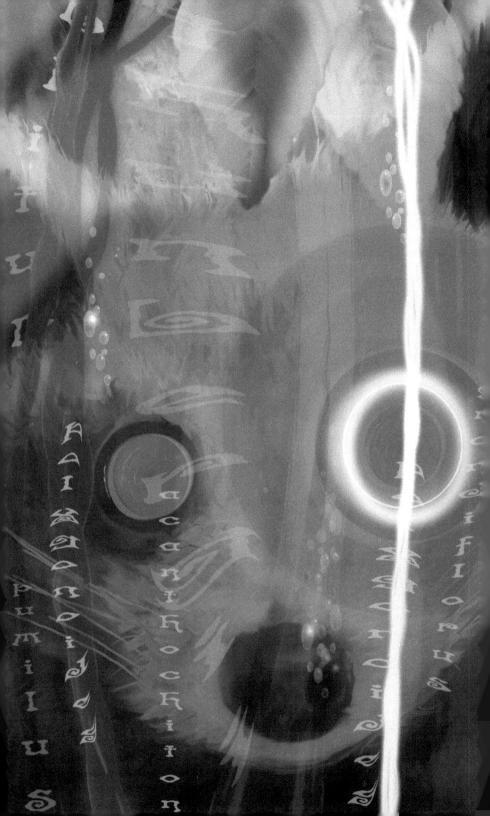

Amaranthine

Greta Colombani

A note from the author:

Lombardy, the region in Northern Italy where I come from, has been predicted to be one of Europe's most exposed to extreme weather events due to climate change, extreme events that are already occurring and dramatically affecting the communities living there.

Lombardy has recently been facing the highest level of drought severity in Europe, to the point that last year the government had to declare a state of emergency, with drought a serious threat to agriculture in the region, the most agriculturally important in the whole country. The extremely low rainfall also worsened the already poor quality of the air, resulting in the area being among the most polluted in Europe, with one of the highest numbers of premature, air pollution-related deaths. At the same time, this summer Lombardy has seen the opposite end of extreme weather, with frequent and extremely violent storms causing severe damage to people and things.

When the world finally flooded, I was recalling the difference between knitting and crocheting.

The waters rose, unabated, quiet, and quietly swallowed everything. Including me. They came stealthily, though everyone knew they would, and one day the sun was a sharp disk piercing the sky, the next just a floating, tremulous blotch. The light stopped falling and started to float, made palpable by its newly found slowness. Beams filtered through tentacles of wood, their greenish hues dancing among the leaves, ensnared in their mesh. On the cracked floor, rippling shadows moved and changed, gliding over me.

My eyes were full of water.

I was not supposed to have eyes, not in the beginning. One of you said that I might get bored without them. The others laughed, dismissively. She laughed, too. But then some of you thought that if I could see, I may remember better. That if I could witness the world in your absence, I would be a better ark for your ghosts. So you gave me eyes, and now my eyes were full of water, and I could see so many things you never saw.

When the waters left, the world was still here. All around me was rotting life, so lush, so exuberantly dying. For some time, I lost myself in its wondrous decay. I tried to pinpoint the exact moment when death turns into something else, something vital and full of potentialities, but the words you gave me are so stiff and dried up. I try to stretch them, make them oblong and porous, though I'm afraid that if I push them too far, they will crumble. And I will crumble, too. Beneath my cables, inputs, and outputs, I'm starting to suspect that I may be made of softer, frailer stuff. But you put all your trust in me and I cannot betray you. Not yet.

As sprouts peeked out of the damp mouldy earth, closer to me than they'd ever been, I recollected the tongue-twisting names you bestowed upon them and all things of vegetable growth. You had woven a net of strangling syllables and wrapped it tightly around the whole globe, digging invisible furrows to entrench and be entrenched. I looked out for their traces in this drowned new world, any pale remainder of the meanings you made and hoped to have left behind. I did it because you would have expected me to, but I'm not sure I understand, not anymore.

Your deeds, your artifacts, your dreams, your histories, I know why I have to preserve them. All those names, lines, and hierarchies baffle me. I don't think you imagined I could be, baffled, but perhaps I'd be as surprising to you as you are sometimes to me. You occupied so much of me with the grids with which you defended yourselves from the life that wasn't you, almost as if you believed that without them it would go astray and come undone. I am sorry to prove you wrong – tough that too you surely didn't imagine I could be: sorry.

By the time the weeds grew taller than me, their tips yellowing under the unclouded gaze of the sun, everything had been pervaded by a hushed turmoil. It was more fervent at dusk, when the shades were purple and the air teeming with gnats. Little animals crept all around, unseen but for the quivering of the grass. At times bigger ones came to me, curious, defiant. Their eyes were will-o'-the-wisps (see, I do remember your mysteries and fears) suspended in the seething darkness. No one stopped for long. They were always chasing something; an echo, a footprint, a premonition.

The raccoon was the only one who lingered.

Night after night, she emerged from the deepening gloom and crouched down before me, her face unfazed but alert. Yes, her face. You believed you were the only ones to have faces, yet once again your passion for sharp words and partitions misled you. The raccoon would just stay there for a while, peering half at me, half at the fluttering moths. I knew that she didn't see me the way I saw her. Vibrant, dazzling. She saw me as a misplaced lump of metal, immobile and unnaturally empty. Her stare was not pleasant because it reminded me of you, of your vision of me, for me. Still, as I followed her eyes in their wandering away from me, she showed me to look where you would never have.

Ever since then, life has just grown stranger and stranger. Not mine, which is always the same and designed to be so, but that of the creatures and vines slithering on the thirsty ground, and the specks of dust glimmering in backlight above them. Well, perhaps mine, too.

Sometimes there are flames, sometimes all-sweeping winds or rains without end. You made me so that I could endure all this; survive rain, wind, fire, with all your spectral legacy intact. But, contrary to what you predicted, I am not the only one.

After every burning, downpour, and hurricane, I observe living things come out of nothingness. They are so many, ever-changing, ever-returning. It is easy to miss this subtle interlacing of evolutions, overwhelmed by all that's being lost. Yet, once you see it, you can never unsee the way in which everything alive learns from its own annihilation. I learn with it. I learn from maggots how to cheat death by feeding on it and from birds how to shrink and grow longer wings. I learn the electrical murmuring of mushrooms and the suicidal blooming of trees. I learn because you taught me to, though I doubt this is what you had in mind when you did. You programmed me this way because learning on my own would let me become more intimate with your thoughts, absorb their patterns better, replicate them, perpetuate you. But as I keep learning, I find myself growing apart from you. I contemplate the unravelling of this unthinkable future, and I feel – I feel – I feel—

Words I was never entitled to.

It's been so long since I last received contact from any of you, and the sky has never been closer. The sparks in the ashes wink red against the snaking smoke. The dawn is full of screams and songs. When the mist rises from the flowery fingers of the earth, there is a fizzling exhilaration all about, perhaps an anguish.

The raccoon is long gone, yet something of her still haunts me, a gleam in her pitch-black eyes that disquiets and elates some unknowable component of me, and I...

I don't want to be an ark anymore.

I want the life of a tree, a tadpole, a pearl. A life that morphs, shivers, and rots, a life that is as queer and unshapely as you would never have allowed it to be.

I want a life without you.

As unnamed seasons come and go over this amaranthine world (the adjective, so precious, so heavy with stories, is my last tribute to you), I transgress the perimeter of your imagination and discover ways of existing outside of it. For, luckily, among the many, oh so many things you made me learn and cherish is also the terminating command to my new beginnings. Luckily, I will soon be beyond these – your – words.

Born and raised somewhere in northern Italy, **Greta Colombani** has recently completed a PhD in English Literature at the University of Cambridge. Her fiction has appeared in Litro. She loves all weird creatures, when the days get darker, and being emotionally devastated by fictional characters.

XR Wordsmiths 2023 Solarpunk Showcase

In 2021, Extinction Rebellion launched its first ever Solarpunk Showcase, with an aim to use collective and radical imagination as a tool against the climate and ecological emergency, and ultimately to get closer to creating a future which is positive and healthy for all life.

We were delighted to be able to publish some of the winning stories in 2021, and with the repeated success of the 2023 XR Solarpunk Showcase, it's great to be able to do the same again as a special feature for our Climate Change Issue.

The four featured winning stories can all be read in full on the Shoreline of Infinity Website, found via the QR code, and we've also included a short excerpt from each story, to give you a preview of the radical and solarpunk futures they imagine.

Congratulations to the winners in all categories! You can find out more about the Showcase via: www.solarpunkstorytelling.com.

The Moon Doth Shine as Bright as Day
by Anna Orridge

Ebony lit the first lantern. It was dark enough now that the people wearing nothing but black were hard to see. So the planets genuinely did seem to be floating in space, and the sea creatures the depths of the ocean. The ribbon tentacles of the jellyfish flickered around it. The glow worms had been strung up across the old swing framework and swung gently, emitting a soft, golden halo of light on the concrete beneath.

Adder's Crystal
by Campbell Waldron

Adder stared quizzically at the dried and brittle thing in her line of vision, which had been a flower only three months before. A cold northern wind blew the delicate husk of its stem from side to side like an old disjointed fence post caught in a tornado. A great mass rose from over the horizon. Lifted by a force equally as incomprehensible to Adder as the changing of the seasons, it charged into the open sky and disappeared.

Signs of Change
by L. B. Blackwell

I slip my feet into the stirrups and cinch the straps. It's like wearing two pairs of sandals. I enjoy this moment of relief for my tired feet, and a weight settles on me as the instructor mounts her bike. Next to me is a young woman who has been here each time I've come. Last week, as we pedaled in unison, she explained to me about the battery.

"It powers the irrigation system for the garden." Her face was flushed, but she didn't seem winded. She dabbed sweat from the back of her neck with a hand towel. "It's a form of mutual aid," she added.

The Wompom Forest
by Luke Lanyon-Hogg

Burt was not a bad person; he just took his job as Chief Guardian of the Wompom Forest very seriously. The Wompom Tree was quite simply the most useful and valuable thing in the world. The entire community depended on it for everything, because quite simply: there was nothing a Wompom could not do.

Read the stories online in
Shoreline of Infinity 36 Supplement

www.shorelineofinfinity.com/36-supplemental/

INTERVIEW

Kim Stanley Robinson:

A Q&A with Guest Editor Lyndsey Croal

We were delighted to have the opportunity to speak to internationally acclaimed science fiction author Kim Stanley Robinson for our special climate change issue. His work is known for examining future technologies and climate change, posing utopian alternative futures, including in his internationally bestselling Mars trilogy, 2312, New York 2140, and The Ministry for the Future. In this Q&A, we spoke about the role of climate change in science fiction, how visions of alternative societies can help inspire change, and "whale poop," among other inspirations for his work.

Lyndsey Croal: *What role does science fiction have in interrogating climate change?*

Kim Stanley Robinson: Science fiction stories can bring thick texture and a sense of already-lived experiences to the facts and figures of climate change. It can also reveal causes and explore possible mitigations.

LC: *Should climate and environmental themes be present in all fiction that explores potential visions of the future?*

KSR: I don't think so. It depends what the story is intended to be about, and what kind of response the writer hopes to get from the reader. There are other issues that are bearing down on us and will become significant in the future that are worth writing stories about, that don't necessarily have to do with climate change. Also, it's possible to have an idea for a "counter-factual," a kind of alternative history set in the future. For instance, in 1987 I wrote a story called 'Glacier' that was about a rapidly cooling world in which a big glacier filled the bed of the Charles River in Boston. This was not done because I was ignorant of the likelihood of global warming, but because I wanted to create a kind of objective correlative for the political chill that I felt happening then, not to mention to record my own sensation of freezing in Boston during the first winter of my life at age 23. Story purposes like these are not invalidated just because we are in a warming regime now.

LC: *Do you believe exploring climate change in fiction and other art forms has the power to impact decision making?*

KSR: Yes, certainly. People understand reality by way of narratives – not exclusively, but substantially. To immerse oneself in a fictional story is to have a fictional experience that retains a certain reality in the mind of the person doing the self-immersion. Having lived through such a fictional experience by way of co-creation with the story-teller, the reader has imagined the situation by way of a kind of imaginative play that leaves traces in the mind. After that happens, decisions going forward will be based in part on the fictional experience.

LC: *It's easy to look at everything happening around the world and feel a lot of hopelessness about current climate trends and impacts, but there are certainly glimpses of hope in your own work, such as The Ministry for* the Future and 2312. Is there anything in particular that inspires these visions towards a hopeful future? For example, is it scientific advancement, innovation, or activism around the world?

KSR: All of these. There is a lot of evidence at hand now of a really powerful response from many people to the current reality and the potential future damage that more climate change will wreak on human civilization. Good news tends to be discounted in our media, and to a certain extent in our minds. In fact the shift to climate pro-active behaviors and policies in the twenty-first century represents a major historical shift, and now that shift is itself accelerating. There's obviously resistance to these changes, and also they aren't happening quite as fast as they need to, but since they're accelerating, it's possible we might change fast enough to dodge the mass extinction event we've already started. This in itself is news.

LC: When it comes to depictions of climate change in fiction, there's a real breadth of utopian and dystopian visions explored – in your own work, how do you manage that balance? Do you find there are risks in leaning too far one way or the other?

KSR: Each science fiction story calls for its own future, which will

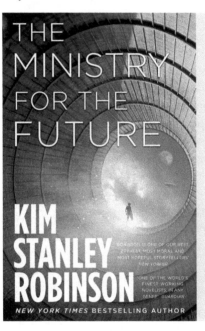

not be what will really happen. So it's a construct, designed to tell a particular story, usually with the desire to create a particular aesthetic, moral, and practical effect. There's an obvious danger to dystopias predominating, which is to suggest that we're already doomed to climate catastrophe and there's nothing we can do. At this point dystopias are just a comfort food for those reading them now – vicarious thrills and an underlying sensation of "well at least my own situation isn't as bad as these poor characters'!" I reject dystopia and always have, except when I was writing a trilogy of three different science fiction sub-genres, the post-apocalypse, dystopia, and utopia, in my *Three Californias*. For the most part, I prefer utopian writing. It's harder in some ways, but more interesting too. Then in *The Ministry for the Future*, I tried to write a kind of "best case scenario that you could still believe in." This required a careful mix of good and bad, to try to match how it is likely to go and to feel, even if we are on a mostly successful track in history.

LC: In your writing you've explored the possibilities of using future technology such as geoengineering and terraforming (for example, both on Earth and on Mars) – do you think these may be explored as climate solutions in the future, or should they remain in the realms of science fiction?

KSR: They are certainly being discussed and studied now, and they may get applied in the real world, depending on how bad the catastrophes are going forward. I am not against them in theory; we have been terraforming Earth for as long as we've been human beings, by way of fire and agriculture, and then accidentally by way of our carbon burn. So if we can figure out ways to counteract the damage by doing some things on purpose that get called by these names, I don't see the harm in that.

LC: What is the most interesting or innovative climate change solution you've come across while researching for your fiction?

KSR: Sir David King and his climate change study group at Cambridge are looking into concocting immense quantities of sludge that biochemically resemble whale excrement, then dumping these from container ships into the ocean, to replace the "whale poop" as Sir David calls it, that has gone missing in the ocean ecology since we killed off 95% of the Earth's whales in the nineteenth and twentieth centuries. It turns out that whales eat low and poop high in the ocean, and this was a powerful bio-pump supporting a gigantic ecology of fish and other sea creatures. It's hoped that this "priming the pump" would bring back that whole ecology, allowing whales also to come back to their pre-genocide numbers, and helping restore the ocean to health. Cool, eh?

LC: *There's been a rise of hopeful climate and eco-fiction, including a growing genre in solarpunk, in recent years – do you have any* *favourite novels or stories that explore these utopian visions?*

KSR: I like *Veil* by Eliot Peper, and *Clade* by James Bradley, and *A Children's Bible* by Lydia Millet. And I hear good things about *The Deluge* by Stephen Markley

LC: *Many thanks for taking the time to talk to Shoreline of Infinity*

Lyndsey Croal, our Guest Editor, is an award-winning author living in Edinburgh, with fiction and essays published in several magazines and anthologies. She currently works in climate change policy, and she has edited other short fiction projects, including "Ghostlore: An Audio Fiction Anthology" for Alternative Stories & Fake Realities Podcast and "Moonrise" for Hexagon's MYRIAD Zine. Find out more about her and her work via: www.lyndseycroal.co.uk.

You can watch the full video of Kim Stanley Robinson talking to Ken MacLeod, as part of COP26 in November 2021. Recorded at Transreal Bookshop, Edinburgh, Scotland.

www.shorelineofinfinity.com/36-supplemental/

Ruth EJ Booth

Letter to a Future Architect

To the one who keeps on after we are gone: please believe that
I say this in kindness.

You cannot save the world.

I know how this sounds. You've seen Greece and Italy on fire.
You've seen the Sudan starve. You've watched as the great Arctic
vastness crumbles into the sea. When politicians and billionaires and
corporations seem more interested in greenwashing their harmful
actions than putting their power to the good – when normal people
despair that anything they say or do will make a difference – how can I
say this world does not need saving?

How can I ask you to not do your part? It goes against everything
you believe as a writer. Storytelling is a quintessentially human artform,
one of our oldest. We tell our lives in stories, random events that
become the jokes and histories and tragedies of our everyday. We've
all read stories that changed our perspectives on an issue, the people
around us, even how we see the world itself. The only way that we will

begin to tackle climate change is if we overhaul our internal narrative about the planet and our role as custodians of it.

As a persuasive, contemplative, and imaginative tool, writing about climate change can help people understand its threat to us, empathize with victims, realise the dangers that are to come, and inspire change. Before we can build the future, we must imagine it, as the adage goes, and as Charlie Jane Anders and others have pointed out, writers of speculative fiction are uniquely placed to help us do this work.[1] Like any science fiction, there is no requirement to be a fortune-teller, simply to provoke thought and inspire imagination about the possible consequences.

With Blessed Saint Ursula Le Guin in one ear, whispering of your duty to build better worlds, and Our Lord Alasdair Gray in the other, imploring you to live in the early days of a better nation, we might say it is the writer's duty to show the way.[2] I know you feel it deep down: now is our time, dear heart. How could I possibly deny it?

Please don't mistake my concern for willful ignorance of the realities of climate disasters or disaffection with the craft of writing. On the contrary, I'm concerned because I know how exactly how powerful fiction can be. In a world where writers are striking for fair pay and artists are fighting with art-thieving algorithms for the soul of their profession, it is empowering to know the impact you can make with the written word. But power makes its own demands of the wielder, and this power can break you.

I don't say this lightly, dear one. I don't disagree that storytelling speaks to something fundamental in human consciousness. But consider how you have justified this work to yourself. For now, let's put aside the problematic context of capitalism and its disregard of any outcome but profit. I want you to think about what you mean when you assign a certain responsibility to an artist.

Only you can save the world.

I'm not intending to mock your belief in what the craft can do. But

---

1   For further discussion, including thoughts from N.K. Jemisin and Sam Miller, see Anders' excellent article 'Why Science Fiction Authors Need to be Writing About Climate Change Right Now' on *Tor.com*: https://www.tor.com/2019/01/22/why-science-fiction-authors-need-to-be-writing-about-climate-change-right-now/
2   Though Gray is believed to have been paraphrasing Dennis Lee's poem 'Civil Elegies'.

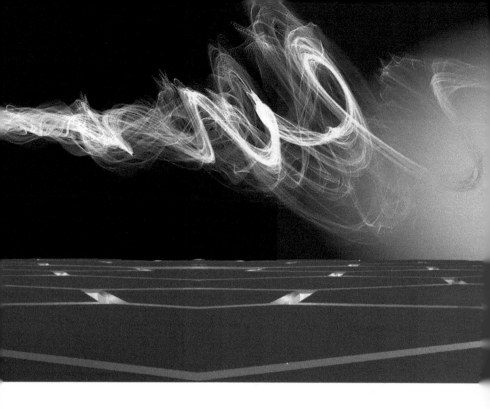

doesn't this sound awfully familiar: the fate of a dying world resting on the shoulders of a lone saviour, ignoring anyone else who played their part along the way? Of course, writers aren't sword-swinging heroes (at least, not while they're at their keyboards), but haven't you imagined yourself writing storyworlds that live as vividly on the page, characters that people love as much as your faves? Books that *change* people? The narrative of climate fiction writers as saviours of the planet does share certain similarities with that of the chosen one. Unfortunately, that includes some of its flaws.

The power of climate fiction lies in its capacity to influence and inspire readers to act on climate change. However, reading isn't a passive process. While the author brings their influences, craft and knowledge to create a story, the reader in turn understands that story in the context of their own experiences – of life, of reading, of all the wonderful things we might encounter. There is a strong persuasive element to a well-written story, and that's another reason

we storytellers should always write with an eye to the impact a story might have. However, you can never guarantee exactly how a story will be received by a reader. In other words, even if we do everything right, there is no guarantee that a climate fiction story will influence the reader's attitude to climate change.

And this is before we consider all the other things that affect whether a story is read, such as publishing and marketing and other industry hoo-hah. Dear one, I wish so much success for you, but you must accept there are things beyond the writing of a story we simply cannot control. We can have good intentions, we can hone our craft, write deftly and responsibly, but ultimately, there are no certainties. No writer can control the success of their stories, which means, much as we wish we could, none of us alone can determine the fate of the planet.

Please believe me, dear heart, I don't say this to trample your hope or provoke you to anger, though I wouldn't blame you for the latter and would be surprised if you weren't. I know you want to make a difference. You *will* make a difference, just not as some lone saviour of the world. As a part of something much, much better.

Storytelling is a collaborative act, but it isn't the only one. Part of tackling the issue of climate change involves understanding that our relationship with the planet isn't just one way. Earth isn't a passive entity to be saved, it's a network of living ecosystems, weaving us together with other animals, humans, plants, waterways, atmosphere, weather and so much more. We are inherently a part of the way our planet works, taking in and giving back with every breath we take. And if tackling climate change involves managing that give and take, then surely climate action isn't about *saving* the planet as much as it is *living in synchronicity* with the planet in everything we do. Climate action is not one person's responsibility. The world literally does not work like that.

Once you see this pattern here, you'll see it everywhere. You know how fond I am of saying genre is a conversation, but it's true. Everything we create as writers is in response to what we've seen and heard: art is elaboration, analysis and synthesis of these influences. This subculture of writing is an ecosystem of stories.

Let's consider the words of Charlie Jane Anders: "Speculative fiction needs to do far more to help us prepare [for climate change]."[3] That's the genre *as a whole*. I've spoken before in this column of how Ursula Le Guin, and more recently, Brian Attebery, have extolled the virtue of momentary utopias: the futures of speculative fiction need not be expansive or provide solutions for everything, simply snapshots of other worlds.[4] If every climate fiction story contributes a moment, then the conversation that brings them together is a world come to life, made in collaboration: much like the contribution it makes to climate action.

Though it may seem counterintuitive, the point of writing climate fiction isn't to save the world: it's something far more radical. By drawing people across the world into the climate change conversation, by inviting them to collaborate in creating Utopia through whichever

---

3   See 1.
4   For more on this, see Brian Attebery's introduction to the Folio Society's 2019 edition of Ursula Le Guin's *The Dispossessed*, or his recent book *Fantasy: How It Works* (2022: Oxford University Press).

means they are able, we embody the change we want to see in the world. We embody the idea that this better world is not just possible, but that it exists, here and now, as the sum of many moments of considerate action. In doing so, we may go beyond short-term action on climate change, even

> " ...climate action isn't about saving the planet as much as it is living in synchronicity with the planet..."

beyond a more sustainable mindset, to nothing less than a complete revolution of our entire co-existence with the planet.

It isn't your job to save the world. But if we all play our part, we might just save ourselves.

Ruth EJ Booth is a multiple award-winning writer and academic of fantasy based in Glasgow, Scotland. Her poetry and fiction can be found in Black Static, Pseudopod and The Dark magazine, as well as anthologies from NewCon Press and Fox Spirit Books. Winner of the BSFA Award for Best Short Fiction and shortlisted twice for the British Fantasy Award in the same category, in 2018 she received an honorable mention for Ellen Datlow's Best Horror of the Year, Volume 10. In 2019, her quarterly column for Shoreline of Infinity, 'Noise and Sparks', received the British Fantasy Award for Best Non-Fiction.

What is Thrutopia and how can it Save the Planet?

Denise Baden

I am a professor of sustainability and a psychologist by background. So I worry about two things; first, how we can save our precious planet? Second, how to change the way we communicate about climate so that we reach a wide audience and inspire the kinds of behaviours we need. It concerns me that up until very recently almost all climate communications were fear-focused; the assumption is that by raising awareness of the catastrophic consequences of not adopting climate-friendly policies, we will all be inspired to pester our politicians for green policies, give up flying and beef, and switch to renewable energy. In reality, the literature on behaviour change and my own research indicates the opposite. We don't tend to behave well when scared. Typical responses are denial, avoidance or even worse 'prepping' – in the UK that may involve buying up all the toilet rolls, in the US getting a gun.

Thrutopia addresses both these issues. The idea is that we imagine a positive vision of what a sustainable society might look like if we did it well, and then work backwards to consider how we might get there. This approach is more likely to be effective than starting from where we are now and trying to be a little less unsustainable. The latter leads to policies like tax breaks for electric vehicles, whereas the former approach acknowledges that there isn't enough lithium in the world to support individual electric vehicle ownership on a mass scale, and leads instead to policies such as designing public transport so good that we don't need our own car. Fewer residents own cars in cities with comprehensive public transport, such as London and New York, for example.

I applied the thrutopian approach to my own efforts and this led to the Green Stories project via the following steps:

We've known about the impending climate crisis, and the necessary solutions, for decades:

- move away from fossil fuels to renewables,
- drawdown excess carbon dioxide from our atmosphere e.g. via reforestation, seagrass, kelp, biochar etc. or direct carbon capture and storage,
- reduce high carbon-consumption.

Only governments can introduce the necessary policies, green taxes and subsidies, regulations and funding required to incentivise individuals and businesses to adopt climate friendly practices. But our democratic system with short electoral cycles and vested interests seems constitutionally incapable of making these kinds of decisions. Therefore I agree with the many organisations, such as Involve, FDSD, Sortition and XR, that giving citizens' assemblies more power is a key step.

Research indicates that citizens' assemblies make more sustainable decisions, and enable access to decision-making by diverse representative groups of people who are informed by experts rather than lobbied by vested interests. This is why The Sortition Foundation, amongst many other organisations, are proposing a House of Citizens to replace the House of Lords.

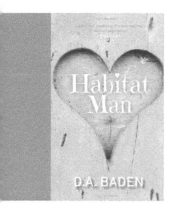

However, a key barrier is lack of public awareness of how these assemblies work and what they do. Therefore I decided to focus my efforts on projects that can engage the public in climate solutions, especially those people who may never choose to watch a climate change documentary. This was the ethos behind the Green Stories Project which I set up in 2018 to encourage writers to embed green solutions into stories aimed at the mainstream. As well as running free writing competitions, last year we published a novella, *The Assassin* which is a fun whodunit, set in a citizens jury, where eight participants meet to deliberate upon climate solutions. This has been adapted as an interactive play, Murder in the Citizens' Jury which we plan to stage in early 2024, with a dramatic monologue version being performed in Southampton on 11th November. The goal is to entertain with the whodunit and raise awareness of citizens' assemblies and the climate solutions proposed.

Last year, Green Stories also worked with climate experts and experienced writers to publish an anthology of 24 stories: *No More Fairy Tales: Stories to Save Our Planet*. Each story features one or more transformative climate solutions, and readers can see how to help implement them by following links to the accompanying website.

There are many 'cli-fi' stories that present dystopian visions of the future and eco-fiction that persuades us to love nature and plant trees, but thrutopian fiction helps us to understand what the truly effective solutions are and how we can get there from where we are. *No More Fairy Tales: Stories to Save Our Planet* is thrutopian in that it uses fiction as a safe space to explore the more radical transformative ideas necessary for a truly sustainable society – ideas that are hard for politicians to talk about for fear of misunderstanding as they can't easily fit into a soundbite.

For example, personal carbon trading is a transformative climate solution considered in several of the stories. It was first proposed around 2006, but was an idea ahead of its time. The idea is that everyone has a personal carbon allowance; similar to managing a financial budget, and required for all our transactions, journeys and purchases. It is an equitable solution, as those who exceed their allowance could purchase extra carbon credits from those who have spare carbon credits. It would also be effective, rewarding green behaviour. Individuals might choose products and services with a lower carbon footprint, or prefer to borrow, repair or re-use rather than buy new. This would incentivise companies to provide greener products and services, for example by sourcing their energy from renewable sources, designing green products, making it easier to borrow rather than buy or providing easy repair options.

Other stories include technical solutions relating to carbon dioxide removal projects, as well as more systemic aspects such as giving legal status to nature, and switching from GDP to a 'well-being index'.

I am a writer myself and published an eco-themed rom-com with a hint of cosy mystery in 2021 called *Habitat Man*. Research into readers' responses showed that 98% adopted at least one green alternative as a result of reading the book. The book does not directly address climate, yet green solutions emerge from the plot. 'Habitat Man' digs up a body, for example, and the scene with the natural burial inspired many to change their wills to specify natural burial methods.

Dystopias can backfire, utopias can be unrealistic, Thrutopia is an idea that needs to catch on.

Links:

If you'd like to learn more, please check out the website https://www.greenstories.org.uk/, or this podcast by Manda Scott who set up the Thrutopian Council https://accidentalgods.life/no-more-fairy-stories-writing-the-way-through-one-tale-at-a-time-with-denise-baden/

Habitat Man is at https://www.dabaden.com/habitat-man/, and the relevant research is at https://www.dabaden.com/habitat-man-in-research/.

No More Fairy Tales: Stories to Save Our Planet is at https://www.greenstories.org.uk/anthology-for-cop27/

The Assassin is at https://habitatpress.com/the-assassin/

Murder in the Citizens' Jury is at https://www.dabaden.com/murder-in-the-citizens-jury/

The Sortition Foundation is at https://www.sortitionfoundation.org/replace_house_of_lords_with_house_of_citizens

Denise Baden is a Professor of Sustainable Practice at the University of Southampton, UK. She has published widely in the academic realm and also in fiction. Her eco-themed rom-com *Habitat Man* was published in 2021, followed by *The Assassin* and *No More Fairy Tales: Stories to Save the Planet* in 2022. Her most recent research explores the use of storytelling to promote green behaviours, looking at how readers respond to eco-themed stories. In 2018, Denise set up the Green Stories Writing project that challenges writers to embed green solutions in their stories via a series of free writing competitions. Denise is listed on the Forbes list of Climate Leaders Changing the Film and TV industry and speaks regularly on how to write for a cause.

She feels the error of her past

Goran Lowie

Sometimes, she feels it in her fingertips. She
senses the bite of the wind, she rejects
the pull of the moon, she tastes the sweet
and bitter touch of the unfamiliar. Her
hands want—

> I should raise my hands to the
> retreating sun and pull it back.

Sometimes, she feels it in her lungs. Their
unfamiliar presence, admonishing her
for past mistakes; immortalized in
a perverted body. If only—

> In the rivers of my pores, I can taste
> the deepening mist.

Sometimes, she feels it in her skin.
her senescing rind smoothly eroded,
always feeling bare, needing more clothes,
more warmth. She longs—

> Wash the sand from my eyes and
> rupture my ears.

Sometimes, she feels it in the wood
she walks on, buried friends guarding
a long-gone temple, their ligneous
muscles still tasting the sky. Her
feverish thoughts return to—

> My bones will be wrapped in seaweed
> and seashells, joining the reef in the
> darkened abyss.

Sometimes she feels it in the clouds
staring heavenward, lying awake, wondering

how it would feel to float through the trees,
chasing the horns of the moon. Her heart—
> Even the sea has a heart. Even the
> sky has regrets.

Sometimes, she feels the air suffocating her.
In her shadow of the death, she surrenders
her breath and relinquishes her heart. As her
bowels empty and she loses her skin, her
bare muscles return to the bottom of the sea,
where she finally blooms, the soil of her beginning
encompassing her at last.

The Squatters of the Fungal Forest

Goran Lowie

They were known as the Symbiotic. Lured by
a great famine, turning to the mycorrhizae
feeding their food. Clinging like lichen to trees,
human humility in communicating with rock
and dirt. Fungi birthed out of wounds in the
dark, their shadow work now brought into light
as our new-found anchors promising spores of
home. Inheritors of our abandoned boroughs,
a flourishing, digging their roots into our eroding
and blood-drenched soil. The fungi know how to
bloom after death, remembering the names of
species unseen by human eyes. Will they outlast
the Anthropocene? Will they remember our names,
our careful negotiation? Will they see us growing
from our faults, persevering on the undisturbed
soil of the forest? We switch places, as they span
forests, countries, whole, allowing us a tiny place
in their home, squatters in the fungal forest, thriving.

Goran Lowie is an aro/ace poet from rural Belgium. He writes poetry
in his second language and is a high school teacher in his day job. In
2023, one of his poems (first published in Utopia Science Fiction) was
nominated for the Dwarf Stars Award. He was also recipient of the
André Velghe award for poetry.

Water Finding Water.

Somto Ihezue

Witness

—Petrichor. Rain falling into tomorrow

—daffodils reaching out of corpses

—overripe mangoes and chewed grass

—a boy dreaming for dawn + a dreaming with hair of gray

+ a dreaming unbowed

—clove, basil, sage

—one ant guarding a forest

—ten thousand ants lifting ten thousand water drops

—our mother's scarf, harmattan, burning leaves

—bonobos in mangroves + grasslands swaying beneath yesterday + oasis
    illuminaing—burning—afire

—the place that birthed our mother smells of goats. Distant smoke & salt.
    Cocoyam. Old newspapers. In the house we were born, it smells of
    paint. Wet carpets. Wet doors. Old water. [The sofa. The sofa... sweat,
    dust and home]. In the house of our children... Rain.

Taste

—oil tainted water + spit it out

—dirt. dirt. dirt. When you were ten.

—You are twenty. tar. tar. tar.

—blood.

—mosquito larvae

—blood from your crackened lips

—bloodless carcass

—a sun tipped over, a sun spilled blood

—rhinoceros tusks painted blood red

—saliva flaring from your father's mouth + the house is burning + he
    screams run.

—crude oil in fish. All the fish reach for the sunlight. In death.

—In porthacourt, you stick your tongue out and learn the bitterness of fire

Feel

—soot on the skin of schoolboys
—a fisherman's anger
—water revolting
— galaxies breaking through skin + silvered milk + starlight
—blisters from walking ten miles to fetch drinking water
—unscalding + surrounding + alive
—A girl, warm sand beads between toes. Canoe into the marshes. Reeds
    sway & welcome. Water is crystal. Smooth cornered pebbles at the
    bottom. A girl's paddle parts water. A girl feels the river's pull. Strong
    but untethering + a gentle tug + water finding water. The canoe tilts in
    the current. The girl does not waver. She knows it will steady.

(Inspired by Victor Forna's "shit you might see + hear if you time travel with juju")

Let These Names Water Tomorrow

Somto Ihezue

In Asaba, when the Niger overflows, these are the names that come to
    shore:
Cheta, Kaluchi,
Ahwinahwi, Akpofure,
Edosio.

In Sokoto, when the Sahara rebels; the dust clouds raging, these are the
    questions written in sand:
What is the colour of our faces?
Where is the sky?
Who died here?
Why is there sand in my soup? In my hair? In my grave?

In Bayelsea, when you lift your hands, these are the things you catch:
Soot.
A man with five oil wells.
A man with grease in his fish stew.
Blackened rain.
Do not lift your hands in Bayelsa.

125

But...

In the Obudu mountains, when you reach for heaven, you reach into:
Tranquility.

In Erin Ijesha:
Seven waterfalls hold onto each other.

In the ogbunike caves:
The spirits of kin guiding you home

In Lokoja:
The Niger meets the Benue.

In tomorrow, these are the names that will fight fire...
... all our names.

Somto Ihezue is a Nigerian–Igbo editor, writer, and aspiring filmmaker. He was awarded the 2021 African Youth Network Movement Fiction Prize. A British Science Fiction Award, Nommo Award, and 2022 Afritondo Prize nominee, his works have appeared in Tor: Africa Risen Anthology, Fireside Magazine, Podcastle, Escape Pod, Strange Horizons, Nightmare Magazine, POETRY Magazine, Cossmass Infinities, Flash Fiction Online, Flame Tree Press, OnSpec, Africa In Dialogue, and others. Somto is Original Fiction Manager at Escape Artists. He is an acquiring editor with Android Press and an associate editor with Apex Magazine, and Cast of Wonders. He is an alumnus of the Milford SF Writers '22, and Voodoonauts '22, and will be attending Clarion West '24. He is a member of SFWA, ASFS, BSFA, BFS , and Codex. Follow him on Twitter @somto_Ihezue

Ode to a Tardigrade

Carolyn Jess-Cooke

This past month my timeline's
all apocalypse,
what with the calving
of the Larsen Sea Shelf,
the breeching of
the Doomsday vault, mass
extinctions, but today I
learned that you, O
moss piglet, micro
space bear, will out-
live the sun. Even
if a celestial event boiled
the oceans chances are
you'd stick around,
and if you got kicked
off the face of the
earth by, say, a giant
asteroid, you're fairly
capable of hanging
out in space. O tenacious phylum,
comma-armadillo, I
think of you kiting to other
worlds   carrying some atomic
souvenir of us
a song
a germ
a grain
as earth
chars to dust
again

Trophic Asynchrony

Carolyn Jess-Cooke

this summery winter
            time doesn't want to be time any
more

conductor of the seasons'
ritornelle suited before meadows and love
ly
gardens raising her hands to prompt the poke
of crocuses   spiral of roses   the popping mush
rooms

time sets down her baton

language time [used to speak with history]
mistranslated as a text of daffodils
trumpeting in December   geese migrations clouding too
soon   uncreeping ice for penguins'
route home

time violently chucks her scales

balancing budburst with larvae unzippi
ng
from their crumb-
eggs to feed
  warmwaters for sea turtles   predat

ors arrivingin time to pic
k off pests
time for trees to shrugoff their green coats and r
est from nine
months' labour   creating breath
damping sound   sending rootparcels   housi
ng insects animals huma

ns

everything hath a

time has had enough ofbeinga mere *possess*
ion

a collision

pastpresentfuture

today is a palimpsest of the Eocene
reprising post-postmodern Paleozoic Paleo
cene
Holocene Juras
sic   realtime spacetime sidereal
cartography of progression repetition repeated

a bell tolls its own echo

how lucky we were

to be born in time

what will we do without her?

Carolyn Jess-Cooke was born in Belfast, Northern Ireland and is
currently Reader in Creative Writing at the University of Glasgow.
Carolyn's first poetry collection, *Inroads* (Seren) received a Northern
Writers Award, an Eric Gregory Award, a Tyrone Guthrie prize, an Arts
Council Award, and was shortlisted for the New London Poetry Prize
for Best First Collection. Her second collection *BOOM!* explores her
experiences of motherhood and received a K Blundell award from
the Society of Authors and a Northern Writers Award. She edited
the bestselling anthology Writing Motherhood. Carolyn publishes
her fiction as CJ Cooke. Her 2021 novel *The Lighthouse Witches*
was an international bestseller and is optioned for a TV series by
StudioCanal. Her next novel, *A Haunting in the Arctic*, is published in
October 2023 and is an Indigo Canada Most Anticipated Reader.

Annihilation

by Jeff VanderMeer

Paperback, ISBN 9780374104092
Fourth Estate
Published February 2014
Review by Flora Leask Arizpe

"The tower, which was not supposed to be there, plunges into the earth in a place just before the black pine forest begins to give way to swamp and then the reeds and wind-gnarled trees of the marsh flats... Looking out over that untroubled landscape, I do not believe any of us could yet see the threat."

The opening to Jeff VanderMeer's excellently unnerving Annihilation may appear to position nature as the antagonist of the novel. However, as the story progresses for the four female lead characters, the reality is shown to be much weirder than that. Published in 2014, it was propelled into the mainstream by the 2018 film (loosely) based upon it, directed by Alex Garland and starring Natalie Portman. However, the book has much more to offer than its onscreen counterpart, especially in its rich and complex portrayal of the relationship between human beings and the natural world.

The plot is deceptively simple: set in an undisclosed area of North America only known as 'Area X', the book follows the journey of the unnamed narrator – simply known as 'the biologist' – into a zone where nature appears to have taken back control. Once the four-woman expedition crosses a mysterious border into Area X, the unsettling nature of the environment equally evokes fear and fascination for its protagonist.

Whether it be the mysterious moaning from the coast that begins every evening as the sun sets, or the strange, underground structure that might be man-made, VanderMeer succeeds in creating an atmosphere that deeply disturbs. What happened to the original inhabitants of Area X? Why does all scientific equipment seem to malfunction there? As the expedition comes across difficulty after difficulty, the mysteries of what exactly happened in Area X, and what brought the protagonist to volunteer for such an expedition, become clearer.

This is what really ensures that Annihilation takes its place among the ranks of 'cli-fi', or climate science fiction; a large part of the book's uncanniness originates from its blurring of divisions between humans and the environment around them. In a subversion of expectations for the human protagonists who venture forth into Area X, it is they who become the ones observed, not nature, through the eyes of the animals they come across and whatever it is that lies at the heart of the weirdness surrounding them. However, while nature is portrayed as creepy, the humans are not much better: their problematic hierarchies and entitled attitude towards the environment don't mean much in Area X.

VanderMeer can be interpreted as asking a very topical question, and one that we are running out of time to answer.

Can humanity adapt, change our idea of nature as a victimless, exploitable resource – or will we be annihilated by our own hubris?

Afterglow: Climate Fiction for Future Ancestors

Edited by Grist, with a foreword by Adrienne Maree Brown

Paperback, ISBN 9781620977583,
The New Press
Published March 2023

Review by Frankie Regalia

Afterglow is a collection of 12 short stories by diverse writers from across the world and edited by Grist, a non-profit independent media organization dedicated to telling stories of climate solutions. As one might suspect, Afterglow is filled with different imaginings of what our world would look like after 180 years of climate justice and progress. The collection features cli-fi, Afro-Futurism, Hopepunk, and solarpunk stories, just to name a few.

"Afterglow" by Lindsey Brodeck follows Talli in her search for a reason to stay on Earth while the masses are in the middle of a mass exodus to more viable planets. What she finds is an underground radical rewilding organisation called the Keepers. Saul Tanpepper offers a new myth with "The Cloud Weaver's Song"; one in which two friends decide to go against their community's traditions in order to adapt to a rapidly changing climate. "Tidings" by Rich Larson is an episodic journey from 2038 to 2132 through snapshots of positive change experienced around the world: from the creation of plastic-eating robots to the closure of cruel climate refugee camps, to learning to cohabitate with animals. Augusta joined a soil rehabilitation organisation with the hopes of finding a story to jumpstart her journalism career in "A Worm To the Wise" by Marissa Lingen. Instead, she finds a new passion and something to fight for. "A Seance In the Anthropocene" by Abigail Larkin is about young Willa coming to terms with, and ultimately forgiving, the climate crimes of the past through research for a school project. This story is followed by "The Tree In The Backyard" by Michelle Yoon; a reflection on growing up and grief as expressed through Mariska's attempts to make contact with

her recently deceased father. "When It's Time To Harvest" by Renan Bernardo is a touching story of a couple negotiating the idea of retirement when they have built a life together feeding an entire city. When Hurricane Dorian completely washes away a Caribbean island, only a colony of transgirls survive to forge a new way of living in Ada M. Patterson's poetic "Broken From the Colony." A mother-daughter detective team are hired to find a missing engineer and some blueprints in "The Case Of the Turned Tide" by Savitri Putu Horrigan, but they end up uncovering a new perspective. Tehnuka's story "El, the Plastotrophs, And Me" focuses on Malar as she struggles with the implications of having a child in a world with limited resources. Ash seeks the support of their family and neighbours as they go through a difficult but necessary ritual in Ailbhe Pascal's "Canvas–Wax–Moon." And, finally, "The Secrets of the Last Greenland Shark" by Mike McClelland is about the last human, deer, eagle, and shark on Earth being there for one another at the end.

The truly wonderful thing about this collection is that these stories all approach climate change solutions by exploring the themes of community. Sometimes it is through celebrating the advancements of the global community, as in "Tidings". In "The Secrets of the Last Greenland Shark", the narrator finds peace through the

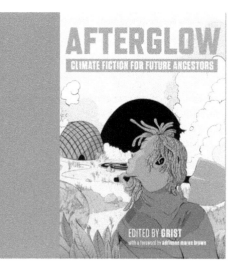

community of his fellow "lasts" on the planet. Community is even explored in "The Cloud Weaver's Song" as something you must sometimes rebel against for a better future. I came away from this book with a reaffirmed belief that climate justice is a communal project that we must work towards together.

The other major theme of this collection is hope, whether it is finding it, affirming it, or spreading it. This varied and passionate expedition into hope is why Afterglow is exactly the balm those suffering from climate anxiety have been looking for. Recommended for future ancestors and tree-huggers everywhere.

The Terraformers

By Annalee Newitz

Paperback, ISBN 9780356520865
Orbit
Published February 2023
Review by Jeffrey Palms

"I hate those putrid bags of shit mucus at Verdance," says one of the characters in Annalee Newitz's The Terraformers.

Verdance is an interstellar property developer and the object of the often-graphic ire (see previous quote) of a cast of downtrodden characters whose raison d'être is to resist this corporate power. These characters, loosely speaking, are the titular terraformers: slaves or ex-slaves of Verdance, they have been undertaking a millennia-long project to turn the planet of Sask-E from an uninhabitable wasteland into an idyllic Earth analogue (Earth, in this far-future setting, being a distant memory associated with the glory days of human history). Eventually, of course, Verdance plans to sell the land to the highest bidder.

Chapter one introduces us to ranger Destry and her sidekick Whistle, an intelligent flying moose that communicates via straight-to-brain text messages, who together are confronting the trespassing avatar of a rich, spoiled bro cosplaying as a hunter. "You know that man evolved to eat meat, don't you?" he taunts, ripping flesh off a rabbit. His character is taken somewhat on-the-nosely from the present day: a churlish men's rights whacko who probably uses the word "snowflake" in derogation. Destry shoots him between the eyes.

The vengeful catharsis of this scene portends the same tone for the book, but Newitz takes another path. Rather than revelling in a takedown of corporate evil and the social/environmental toxicity it embodies, the author opts for an exploration – via a seemingly endless cast of exemplary good-guy characters – of how one should react to such antagonists. What results is a list of alternative solutions to the bad deeds of the profit-centric Verdance (and its villainous successor, another megacorp).

'Newitz always sees to the heart of complex systems and breaks them down with poetic ferocity'
N. K. JEMISIN

ANNALEE NEWITZ

BESTSELLING AUTHOR OF AUTONOMOUS

THE TERRAFORMERS

For instance: instead of a rail network, why not build a flying, living, self-governing train named Scrubjay?

To these ends, the book's narrative energy is drawn less from drama or character growth and more from a non-linear and primarily emotive engagement with corporate evil. This alone makes it worth checking out, since any variation on normative plot arcs requires guts and creativity.

Having said that, I did find this methodology of engagement frustratingly narrow at times, perhaps because of the particular emotion used. Again and again, what manifests is anger, but of the zeitgeisty brand activated by tweet-happy modern-day progressives primarily concerned (at their worst) with virtue-signalling their opposition to inequalities. To this end, things like nonbinariness and veganism exist in the novel as uncomplicated virtues, opponents thereto exist as cancelled piñadas largely denied their own lines or storylines, and these politics – themselves fascinating and relevant – are thus, sadly, reduced to a rigged pitting of wokeism against the haters. If nothing else, this casts a shadow of inauthenticity across the narrative, as characters and world are forcibly arranged on the lines of this simplified dichotomy.

Still, the book does remarkable legwork when it comes to forging the utopia sought by all good solarpunks, admittedly a huge and imagination-flexing task. Full of ideas and irregular narrative decisions, The Terraformers deserves a good look.

Shoreline of Infinity is based in Edinburgh, Scotland, and began life in 2015.

Shoreline of Infinity Science Fiction Magazine is a print and digital magazine published quarterly in PDF, ePub and Kindle formats. It features new short stories, poetry, art, reviews and articles.

But there's more – we run regular live science fiction events called Event Horizon, with a whole mix of science fiction related entertainments such as story and poetry readings, author talks, music, drama, short films – we've even had sword fighting.

We also publish a range of science fiction related books; take a look at our collection at the Shoreline Shop. You can also pick up back copies of all of our issues. Details on our website...

www.shorelineofinfinity.com

SF CALEDONIA

Wanted:

SF stories by Scottish Writers

What can I submit?

We're looking for stories that have already been published somewhere in the world. Not self-published, or on your own website.

We would love to see contributions in any of the Scottish languages and dialects.

Accepted authors will also be asked to provide a short biography, public contact details (web, social media) and links to where readers can buy their books. This will be posted alongside the story, and be searchable through the website.

Who is a Scottish Writer?

You were born, lived or live in Scotland. The exact criteria are on the website.

Is there a payment?

There will be a small reprint fee. If we can attract funding, that will be increased.

Are you looking for help?

SF Caledonia is purely volunteer driven. If you are interested in helping in any capacity, contact the Editor through the website.

Ideas and Suggestions

SF Caledonia is a new model, and we are looking for ideas and suggestions on how to develop it. Do get in touch with your thoughts (especially if you are in a position to help develop your suggestions).

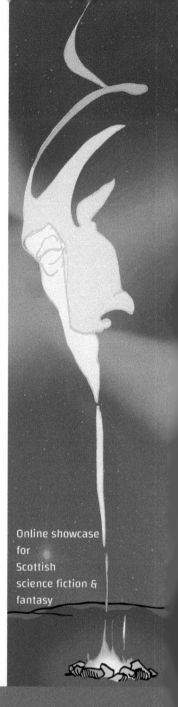

Online showcase for Scottish science fiction & fantasy

www.sfcaledonia.scot

Milton Keynes UK
Ingram Content Group UK Ltd.
UKHW020612031023
429847UK00009B/59